GRASSES - BENEFITS, DIVERSITIES AND FUNCTIONAL ROLES

Edited by **Amjad Almusaed**
and **Sammera Mohamed Salih Al-Samaraee**

Grasses - Benefits, Diversities and Functional Roles
http://dx.doi.org/10.5772/65845
Edited by Amjad Almusaed and Sammera Mohamed Salih Al-Samaraee

Contributors

Suarau Oshunsanya, OrevaOghene Aliku, Victor Lionel Finot, Alicia Marticorena, Romina Muñoz, Roberto Ridríguez, Katherine Dunster, Yusuf Bozkurt, İlker Yavaş, Aziz Gül, Beytullah Ahmet Balcı, Nurdan Coşkun Çetin, Rafael Goulart Machado, Abdelhak Maghchiche, Phool Shahzadi, Amjad Zaki Almusaed

Published by InTech
Janeza Trdine 9, 51000 Rijeka, Croatia

Publishing Process Manager Maja Bozicevic
Technical Editor SPi Global
Cover InTech Design team

First published September, 2017
Printed in Croatia
Legal deposit, Croatia: National and University Library in Zagreb

Additional hard copies can be obtained from orders@intechopen.com

Grasses - Benefits, Diversities and Functional Roles, Edited by Amjad Almusaed and Sammera Mohamed Salih Al-Samaraee
p. cm.
Print ISBN 978-953-51-3493-0
Online ISBN 978-953-51-3494-7

For EU product safety concerns:
IN TECH d.o.o., Prolaz Marije Krucifikse Kozulić 3, 51000 Rijeka, Croatia,
info@intechopen.com or visit our website at intechopen.com.

Meet the editors

Editor, Associate Professor Dr. Arch. Amjad Almusaed was born on January 15, 1967. He holds a PhD degree in Architecture (Environmental Design) from "Ion Mincu" University, Bucharest, Romania. He followed a post-doctoral degree research in 2004 on the sustainable and bioclimatic houses, from the School of Architecture in Aarhus, Denmark. Dr. Almusaed has more than 27 years of experience in sustainability in architecture and landscape with innovative orientation. He has carried out a lot of research and technical survey works and has performed several studies, in the area mentioned above. He is an active member of many international architectural associations. He published many papers, articles, types of research and books, in different languages.

Co-Editor, Assistant Professor Sammera Mohamed Salih Al-Samaraee, PhD, was born in Basra on June 12, 1962. She received her doctorate degree in Medicinal Plants from Basra University in 2004. She is affiliated with Basra University since 1986. Her research areas include horticultural, sustainable landscape areas upon residential, educational and health programs specifically for public spaces, where she published many studies and articles in this domain. She is a supervisor for multiple students involved in master and PhD degree programs. Sammera is a pioneer in her field and has an extensive cooperation between her department and the department of architecture. She is an active member of many local and international organizations.

Contents

Preface

Today, human beings have turned to the profound meaning of green areas in their lives, and it is clear that grasses are a symbol of life, and their importance corresponds to the earliest evidence of people on earth. Grasses have existed on the earth for much longer than people have, and human beings have been the fortunate member of this partnership. Savannah and Steppe form the largest natural grassy areas from different climate zones that are not suitable for forests. In addition, grasses create cultural landscapes, especially in Central Europe. They have arisen in a long process since the Ice Age, where forests have been transformed into meadows and pastures. Appropriately, the earth's first ages of grasses came even earlier than the first age of the grazing animal. Indeed, the word 'graze' was derived from 'grass'. During Miocene times, believed to have begun some 30 million years ago, the forebears of grass came out of tropical forests and swamplands into the cooler and drier lands.

The aim of this book is to embody the essential components, the role and the function of grasses as natural material in our lives. The book will discuss the importance of using different species of grasses in landscape and medicine. The importance of the book is manifested by subject debates, which display multiple viewpoints on how grasses are used in general, and their impact on our lives in particular. The book is divided into four parts and eight chapters.

Section 1, 'Overview on Grass Topic', is an introductory section, which presents a snapshot of the topic through a general reading.

Section 2, 'Grasses in a Complex Agronomic Process: The Benefit and Importance', is divided into two chapters: the first, 'Analytical Interpretation of the Beneficial Interaction Between Microorganisms and Grasses', debates the meaning, benefit and importance of soil microorganisms, the mechanisms of growth promotion of grasses and the main mechanisms of growth promotion of grasses by soil microorganisms. The second chapter, 'Importance of Grass Carp (*Ctenopharyngodon idella*) for Controlling of Aquatic Vegetation', debates the concept of aquatic plants, where grass carp becomes a tool to control aquatic vegetation. This chapter aims to analyse the use of grass carp to control aquatic vegetation.

Section 3, 'Different Grass Species from Diverse World Regions', is divided into three chapters. The first chapter, 'Study of Some North African Grasses (*Ampelodesma mauritanica* and Esparto Grass)', discusses the subject of *Ampelodesma mauritanica*, which is a plant of the family Poaceae from northern and southern Africa. Esparto grass is another topic of this chapter; both are cellulose-based fibres extracted using a basic procedure to remove noncellulosic substances such as pectin, lignin and hemicelluloses. The authors attempt to characterise the Esparto grass fibre obtained from chemical extraction using sodium hydroxide followed by sodium hypochlorite as a bleaching agent. The second chapter, 'Endemic Species of the Family Poaceae in Chile: Taxonomy, Distribution and Conservation', explains how endemic species with a geographic distribution are restricted to a single area and could be especially vulnerable. The third chapter, 'Beyond Turf and Lawn: Poaceae in This Age of Climate Change', debates the traditional uses of Poaceae in various cultures, where many unrealized needs for food, medicine and other material goods could be met elsewhere with knowledge transfer.

Section 4, 'Applied Circumstance of Grass Categories', is divided into two chapters. The first, 'Lemon Grass (*Cymbopogon citratus*)', debates lemongrass citral, which is an essential oil with therapeutic and many other useful effects. The second, 'Vetiver Grass: A Tool for Sustainable Agriculture', debates several applications of vetiver grass, its impacts and resultant benefits as a technology that could enhance sustainable agricultural development. Soil conservation, amendment and stabilisation are the main areas discussed in this chapter, along with crop productivity.

Many propositions, items and significant factors are combined in this book. A current objective of our world is to make the use of green areas more practical and environmentally friendly. Today, the practice is to establish a competent way to use grasses in their original form, starting from their use as food or medical matter to an efficient establishment of grasses in biophilic design in our architectural and city arrangements. This book makes available the theoretical basis and practical application of how a broad usage of grasses can be replicated in our life based on practical ways to give worked-out examples of such problems regarding applied usage. I must acknowledge that it has been a rare pleasure for me to be an editor associated with InTechOpen Publisher. I would like to take this opportunity to thank Ms. Maja Bozicevic, InTech's Publishing Process Manager, and I shall avail this chance to extend my sincere gratitude for her help and cooperation at various phases of the publication process, as well as the researchers who performed experiments and reported their findings.

Associate Professor Dr. Arch. Amjad Almusaed
University of Basra
Basra, Iraq

Assistant Professor Sammera Mohamed Salih Al-Samaraee
University of Basra
Department of Horticulture, College of Agriculture
Basra, Iraq

Section 1

Overview on Grass Topic

Chapter 1

Introductory Chapter: Overview on Grass Topic

Amjad Almusaed

Additional information is available at the end of the chapter

http://dx.doi.org/10.5772/intechopen.70407

1. Introduction

A regulate analysis of the connotation of the word "environment" in the sustainability explains clear circumstances of a being or thing (social, economic, and physical). Moreover, the meaning covers the systematization of the environment upon the physical process. The environment in a broad sense also contains all the natural and artificial factors of the physical, chemical, biological, and social nature in which a human being is a factor motor of community that develops. The importance of maintaining a steady state, the human relationship, and its living environment requires an ability to control the environment in an optimal arrangement of ecological balance. For creating a competent urban zoning, it is necessary to state that green areas in general and grass in special require a delimitation based on a unitary structure of territory structure organized. It has to be created by successive stages (valuable parts), which in the functional aspect consists of areas characteristic of the dominant aspect, the weight of one of the functions, or a greater diversity of the functions, mono-functional multifunctional space [1]. Grasses in the planted area (as a functional urban area) are shaped to serve the specific areas for beauty or recreational purpose. The grasslands have an esthetic function, which can be presented by

- An essential element to highlight architectural objects
- A significant decorative value (by shape or color)

The grasses are the most valuable horizontal green plan. The family itself is the source of the grains that are the stuff of life for most of the humankind. The beauty of grasses is quite different from another planting area [2]. The simplest definition of "the grassland" represents an area of land covered by herbaceous vegetation, consisting mainly of grasses, which is subject to care work and is designed to perform certain decorative, recreational, or sanitary functions. The lawn is the main fund of green space (75–80% of the area; rocks, shrubs, floral arrangements, and massive trees) or naevi (rocks, statuary groups, architectural elements, buildings, and water Lucius), uniting the diversity of the green space. The turf is an item of exceptional decorative value, imparting an atmosphere of calm and tranquility; due to the green tones, the vast stretches of lawn introduce into the landscape a note of solemnity and romanticism. Air temperature is influenced by lawn surfaces; on high-temperature days, the lawn heats less compared to coated surfaces with sand, gravel, brick, asphalt, or concrete. For example, if the average air temperature from the lawn surface is 22–24°C, then the surface temperature of the concrete surface is 30°C, and in the case of asphalt, it is approx. 45°C. Sanitary qualities also consist of air purification by retaining dust and oxygen enrichment. According to destination, composition, and maintenance, several uses of grasses are distinguished:

- For lawns
- For sports grounds
- Rustic

The grass family (Poaceae) is a plant family in the grassland. With about 11,000 species in more than 650 types, it is one of the largest families of the flower plants. It is widespread in all the world's climate zones. Many species of grasses belong to the oldest useful plants and have been of vital importance to human beings since the dawn of the day [2]. All cereals, such as wheat, rye, barley, oats, millet, maize, and rice, are included in this plant group. Landscape forms such as meadows, exuberances, steppes, and savannas characterize vast areas of the world's landscapes. Although the different species in the grass family in many ways resemble each other, there is a great variety. This kind of grasses is no wonder because they are among the plant groups that can easily adapt. They grow in polar regions and deserts, in tropical rainforests, and on windy slopes of mountains. Whole vegetation areas, such as steppe, llanos, prairie, and savannah, are dominated by grass. Most grasses also have the advantage that the stem if bowed by the wind or threaded can rise by firing faster on the side facing the ground. Grasses occur throughout the world and in almost all vegetation types. There are single or perennial plants with hollow stems (straw), which are however compact and often swollen by the leaf feathers (knees). The leaves are more or less ribbon shaped and form a staggered leaf sheath in the lower part. A piggy bank is made up of toroidal, dandelion-shaped leaves (after) and contains one or more flower without a true blister. In many grassy areas, natural fires play a crucial role. Lightning ignites, at the end of the growth period, the dead plant material. The inorganic nutrients in the ash promote the new growth of plants as fertilizers. Also, tree growth is destroyed. The fires thus contribute to keeping the grassland open. Grazing areas inhabit and nurture an ancient and diverse animal world of insects (termites and ants), spiders, birds, small mammals, and not least grazing flocks of large mammals as on the African savannas. Some grasses are sparsely built and are only a few centimeters broad. One- and two-year species usually have single or few shoots in loose leaves with soft leaves. In these grasses, all or most stems bear inflorescences. In most cases, the long-lived species form firmer straw and longer leaves and form a larger or smaller number of nonflowering shoots in addition to flowering shoots. They grow in loose or dense gardens or lawns. The recent growth form occurs when the plants propagate with more or less long overground, creeping, greenish, or reddish stalks, called pricks or using subterranean, white, or brown rhizomes. In addition to the color, the two types also distinguish between the knees on the runners with roots: stools have full leaves on each knee, whereas rhizomes in these places have only small, thin skeletal low leaves. In this way, due to the closely compacted shoots, the typical tube-shaped growth form of many kinds of grass occurs. Most grasses have wood nuts. They do not form a principal or pedestal. At the root of the stem and the knees of the outriggers, many roots are formed, each of which can develop leaflets of first and second order. In this way, root systems with a considerable size can occur. Thus, a single plant of Red Swing may spread 250 m in diameter. The grass is the largest family of flowering plants, and of all vegetative families, it is the most appreciated to human beings. It has an extensive range of uses. However, it is only about relatively few genera that are important as useful plants for humans. Only about 15 genera (it is barely 2%, although bamboo genera are not included) play a bigger role. It is a generic term for Gramineae monocotyledonous green plants where it is found on the plain, hill, and mountain. At present, about 4500 species of the Gramineae, or true grasses, have been classified. Approximately 1500 of these species are believed to be native to North America. The grasses take on more than a fifth of the earth's plant cover; it is composed of plants of small height and produces seeds and is used as a diet for ruminant animals. Varieties of grasses that grow very often and of low heights are used, under the name of law, to cover the green spaces. Grasses carpet the baby's playpen, blanket the old man's grave, and cover somewhere near a fourth of all the other land on the earth. The known species range in size from the miniature "three lawns," smaller than a shirt pin, to the giant bamboo that towers above medium night forest trees. The progress of the grasses is spotlighting the ancient strongholds of the Gramineae. These include the ancient pastures of Western Europe, the principal river valleys of Eastern Europe, the vast steppes of the Russia, the hard used herd lands of Manchuria and the Mongols, in South African veldt, the deep-soiled pampas of Argentina, the campus of Brazil, the llanos of the Orinoco Basin, and the great savannas of Australia and New Zealand. However, the North American Great Plains, for the most part, west of longitude 98° W, keeps their place as the most productive of all the high grasslands of the earth. Currently, grasses are reappearing and otherwise gaining ground in the tropics, the original homeland of grasses where the greatest number of species in any area is still to be found. Grasses are pushing farther and farther up and down toward the poles, invading tundra, following in the wake of sledge runners and the first ruts made by the wheels of planes, tractors, or the wide-ranging jeep. Grasses keep venturing farther into deserts, higher up mountainsides, and closer to ocean edges, but now, as perhaps through hundreds of centuries, the grasses show their fastest increase in the temperate zones. Grasslands from the largest natural grassy areas on earth in climate zones are not suitable for forests. Also, they create cultural landscapes, especially in Central Europe. They have arisen in a long process after the Ice Age, where forests have been transformed into meadows and pastures.

2. Grasses upon human history

Grasses have lived on the earth a great deal longer than people have. In this way, man has been the fortunate member of the partnership. Through apparently hundreds of centuries, man followed the grasses to live on the grazing animals and birds, which they sustained. Appropriately, the earth's first ages of grasses came even earlier than the first age of the grazing animal. Indeed, the word "graze" was derived from "grass." Supposedly, during Miocene times, believed to have begun some 30 million years ago, the forebears of grass came out of tropical forests and swamplands onto the cooler and drier lands. The succeeding Pliocene epoch, believed to have ended a million or so years ago, was distinctively an age of grasses and grazers. During its centuries, both grasses and trees tended to lose their gigantism, while animal life acquired some part of the genetic bigness, which vegetation was losing. Mastodons, giant rhinoceroses, and big horses grazed on the far-spread grasslands of the continents. After the Pliocene, the Ice Age began. Gigantic mountains of ice locked in so much of the earth's water that the Bering Strait, the Isthmus of Panama, and other land links previously submerged were changed into open bridges, which permitted ready intercontinental migrations of grazing animals. The masses of ice that advanced and retreated four times in our present time are described as a late phase of the fourth recession. However, ice masses or no, animal life survived and increased, where and as grasses permitted. Buffaloes, elk, antelope, and deer became the main American grazers on the grassy fringes of the ice fields. Africa and lower Asia stayed free of the ice and became refuges for some of the bigger grazing animals. Supposedly after the last great ice sheets at least partially withdrew, Asians came across the Bering Strait and poured down and across what was by then one of the biggest areas of grass on earth, spreading intermittently from what is now upper Canada to the plains and hills of Patagonia. For many centuries, those earlier Americans sought, slaughtered, and lived with animals, which lived on the grasses. Eventually in the Western Hemisphere, as apparently, he had learned even earlier in the Eastern Hemisphere, man began adapting what he deemed the best of the grazing animals to herd livestock. The North United States of America does not lead the world in either the selective breeding of grasses or the production of nutrients per acre of grasses harvested. In both attainments, other countries certainly France, the British Isles, Australia, New Zealand, and quite probably the Netherlands are still ahead of us. Soviet Russia is claiming the world leader in the genetic development of perennial grain grasses, including wheat and rye, that is, Russian plant geneticists are seeking to turn into biennials wheat, rye, and perhaps other annual grains, and eventually to make them (they hope) perennials. However, in the total value of grass harvests, the United States not only leads but also overwhelms.

3. Grass functions and roles

The grasses are not only the most common of the flowering plant families but also the most important, where some of it represent the basis of our food. These include the four bowls of cereal (wheat, barley, rye, and oats), also corn, rice, millet, etc., and grass seeds for bread, porridge, beer, etc. Grass family has been the main source of food for people in almost all cultures since the early start of the drugstore and also an important source as cattle fodder. Grass has many functions, starting from the medical and therapeutic function to leisurely utility. In the following subdivisions, some of the essential functions of grasses will be represented.

3.1. Therapeutic function

Grasses are one of the oldest therapeutic elements and are found in all continents and cultures where plant medicine also covers parts of pharmacology, pharmaceutics, and toxicology. The herbal medicine is partly based on traditional medicine. Experienced values, traditional knowledge, and traditions play a major role. In plant medicine, only whole plants or parts of plants (flowers, leaves, seeds, bark, and roots) are used, but no isolated individual substances are used. The active components of medicinal plants are subject to natural fluctuations caused by the climate, location, and harvesting time of the plant.

3.2. Protecting and improving the environmental function

The environmental protection and improvement functions are multiple, reaping more or less at different green spaces.

1. The hydrological function is the first function, which is provided by all types of green spaces, being expressed by the greater or lesser capacity of precipitation water to be retained, and to the release of either the atmosphere in the form of vapors or the soil through its percolation phenomenon toward the horizons.
2. The ground protection is an important function that is noticeable in areas lacking vegetation or grasses where the erosion processes are visible. Through the system of rooting, which is a biological soil armature, by the phenomenon of attenuation of the mechanical effect of raindrops, exerted by the foliage of the grasses, which is often arranged in several layers, the substrate constituting the support of the vegetation presents a mechanical, physical, and chemical stability, greatly enchancing the prevention of soil erosion and landslides [3].
3. The climate protection is the main function of green elements, which is exercised by all green spaces under different aspects:
    i. Moderation of the amplitudes and thermal variations
    ii. Decreasing the wind speed
    iii. Improving the humidity of the air
    iv. Improving the intensity of solar radiation

The modeling of diurnal and seasonal variations of vegetation is exercised by the shading effect, through the evapotranspiration processes, by the specific albedo, or by the reduction of the wind cooling effect. The direct effect of grasses is evident in our life where a 1.5 m^2 of uncut grass produces enough oxygen per year to supply one person with his or her yearly oxygen intake requirement [2]. Woody vegetation, through its microclimate, moderates excessive temperatures, so temperatures in hot summer days are lower in the masses of trees, protection curtains, stripes planted alongside the streets, or near them, and in winter, temperatures are higher due to the air movement diminishing the effect.

For example, if the air temperature at the surface of the concrete is 30°C, under the same conditions, the surface temperature of the asphalt will be 45°C, and the air temperature at the grass level will be between 22 and 24°C.

3.3. Antipollution function (healthy)

This function is fulfilled by green spaces, regardless of the size or nature of their vegetation, under different aspects:

- Reducing the physical pollution of the atmosphere
- Reducing the chemical contamination of the air
- Reducing noise effectually

The function of the reduction of environmental pollution can be achieved precisely by the ability of the vegetation to retain, fix, and sediment particles suspended in the atmosphere, fine powders, or smoke. Through the foliage and crown texture of different woody species or the texture of the various grassy vegetation areas, by decreasing the air velocity, the vegetation retains substantial amounts of particles, which are subsequently entrained by the water from the precipitations at the ground level [4]. The physical purification capacity depends on the species, the leaf size, the porosity, the life of the leaves, and so on. For example, an ingenious surface retains 3–6 times more dust and particles. Solids than a mud surface and a medium-sized tree retain 10 times more impurities than the surface of its crown projection covered with lawn. One hectare of oak forest can hold 68 tons of solid particles and dust, for spruce approx. 30 t/ha, silvery pine approx. 35 t/ha, and lime lump approx. 42 t/ha. The reduction of chemical pollution is achieved by green spaces primarily by CO_2 consumption and O_2 produced by the actual fixing of toxic gases resulting from various activities: fuel combustion, chemical industry, metallurgy, oil processing, mineral processing, car, air, or other current activities of the population [5]. Green spaces behave like true biological filters that improve the air qualities, due to the ability to fix through the metabolism of various harmful gases in the atmosphere. Phonic pollution occurs because of various daily activities within or outside localities, activities that generate noise with varying intensities and frequencies. Noises can be mitigated by the dense foliage of grasses, arranged in the form of lanes along the roadways or highways passing through inhabited areas. In cities, street grasses and plantations and rare plantations between buildings and small squares reduce very little noise (only 4–5 decibels) with an unseasonable effect. To obtain the maximum sound effect, they usually combine different relief patterns or different panels with sound effects, with plantations arranged in some devices to absorb and dissipate sound waves [6].

3.4. Recreational function

Selection of particular grasses for landscape areas plays a major role in the activation of recreation function throughout the walking and visiting of some picturesque regions, historic gardens. Recreation function is an organic and spiritual necessity for human beings to escape from the artificial environment of the cities [7]. Recreation can be defined as an activity practiced by a man in his right, in the spheres of culture, art, sports, entertainment, tourism, being an element compensating working conditions, physical, intellectual, or psychological demands to which the person in daily life is subjected in general. The higher these requirements, the greater the need for man to escape from the daily, most of the destinations, in this sense being vegetation areas, urban, or extra-urban green spaces.

The basic tasks of recreation are:

- Relaxation, which removes psychic and nerve trauma caused by tensions or eliminates the temporary fatigue resulting from the daily activity program
- Entertainment or amusement, which removes boredom or the effect of daily automatisms
- The escape through which the individual comes out of his or her usual environment
- Developing personality by which the individual releases for a period the daily automatisms, having creative and innovative behavior and activities

Most of the environments in which contemporary man lives are strong anthropic and often lacking in the conditions of nature. Recreation in nature is increasingly adopted and preferred by the modern person in the urban environment due to the heavily artificial environment in which he operates, but also because of the physical and mental pressure he is subjected to in the various daily situations.

3.5. The function of efficacy

In an exceptional situation, certain grasses can be selected to protect the specific objectives, hydrological resources, and different categories of soil. Thus, some green open space areas have to be provided with special grasses to reduce the spread of harmful substances, open water basins (water accumulations), and drinking water supply installations with a sanitary curtain.

In some other cases, it is necessary to provide around roads regions with special plants that ensure the consolidation of the land or the traffic security (vegetal barriers separating the traffic directions, barriers against the wind, and parasitic effect) [8] by implementing decommissioning projects for industrial and rendering enterprises in the use of the respective territories, by greening the spaces and by landscaping designed to mitigate the visual impact of the installations and the integration of certain specific grasses uses.

Author details

Amjad Almusaed

Address all correspondence to: a.amjad@archcrea-institute.org

Environmental Design, University of Basra, Basra, Iraq

References

[1] Jörg Mildenberger X. Anton Trutmann's Arzneibuch. Würzburg: Teil II; 1997

[2] Almusaed A. Biophilic and Bioclimatic Architecture. Analytical Therapy for the Next Generation of Passive Sustainable Architecture. London, Dordrecht, Heidelberg, New York: Springer; 2010. p. 67, 238, 178

[3] Almusaed A, Almsaad A. Building materials in eco-energy houses from Iraq and Iran. Case Studies in Construction Materials. Elsevier; 2015

[4] Almssad A, Almusaed A. Environmental reply to vernacular habitat conformation from vast areas of Scandinavia. Renewable and Sustainable Energy Reviews. August 2015;48:825-834

[5] Almusaed A, Almsaad A. Urban biophilic theories upon reconstructions process for Basrah City in Iraq. In: Passive and low energy architecture Conference, PLEA 2014; Ahmadabad, India. 2014

[6] Almusaed A, Almsaad A. Biophilic architecture, the concept of healthy sustainable architecture. In: The 23th Conference on Passive and Low Energy Architecture, PLEA 2006; Geneva, Switzerland. September 2006

[7] Al-samaraee SMS. Effect of soil texture and salinity of irrigation water for growth and active ingredients of the henna plant *Lawsonia inermis*. Thi-Qar University Journal for Agricultural Researches. ISSN: 22225005. 2012;1:2222-5005

[8] Al-samaraee S, Hassan A, Alshwally A. Effect of spraying yeast suspension, and time of cutting on growth and content of henna plant from Tannins and Lawsone pigment Lawsonia. Journal of Basrah Researches (Sciences). ISSN: 18172695. 2011;37(5B):104-115

Section 2

Grasses in a Complex Agronomic Process: The Benefit and Importance

Chapter 2

Analytical Interpretation of the Beneficial Interaction Between Microorganisms and Grasses

Rafael Goulart Machado

Additional information is available at the end of the chapter

http://dx.doi.org/10.5772/intechopen.69272

Abstract

Soil microorganisms compose ¼ of the biodiversity of our planet and are responsible for important processes such as the decomposition of organic residues and the transformation of the nutrients contained in these residues into nutrients for plants. The microorganisms also aid the grasses implantation, increasing the grasses yield by means of several mechanisms of plant growth promotion. These mechanisms of growth promotion of grasses can be direct, or indirect. In this chapter, we discuss the main mechanisms of growth promotion of grasses by soil microorganisms. It will be explained how the microorganisms in the soil act favoring the growth and development of cultivated grasses. For this, there will be clarified the importance of soil microorganisms in nutrient cycling, the mechanisms of nutrient capture, the production of phytostimulant substances by microorganisms, and the mechanisms of soil pathogen suppression.

Keywords: nutrient cycling, biological nitrogen fixation, interaction between plants, microorganisms

1. Introduction

The microbial population inhabiting the rhizosphere consists of a wide range of organisms, which together interact directly and indirectly with the cultivated plants. Only with regard to the number of bacteria, it is estimated that there are about 2 billion cells per gram of soil [1]. These microorganisms become interesting to the human species, as they interfere in the yield of the cultivated plants, by means of several mechanisms.

A microorganism is considered a plant growth promoter when it is capable of increasing the yield of the crops of interest. To measure this capacity, the interaction of a given microorganism with some plants of interest must first be evaluated under axenic conditions and in comparison with cultivated plants. It is fundamental that this initial stage is studied in isolation of the interaction of the plant with the organism, thus isolating the interaction of other factors such as climate, environment, and other edaphic or epiedaphic macro or microorganisms to make sure that the effect on the yield of the plants of interest is solely and exclusively due to the inoculated microorganism. Without this initial screening under axenic conditions, it would be impossible to certify and prove that the positive effect observed in the studied plant is due to the microorganism of interest.

Only after the positive effect of the microorganism on the plant has been proven, this interaction will be tested under conditions of greater interference, such as greenhouse, fertilization, or soil conditions with an original field microbial population (nonsterile soil). Under these conditions, the resistance of the interaction to various interference factors will be tested. Once approved in tests conducted under controlled conditions, the microorganisms are tested under field conditions.

Several mechanisms are the mechanisms by which microorganisms act on the yield of plants and can act directly through the production of hormones [2] or nutrient supply, such as nitrogen [3], or indirectly by the suppression of pathogens [4]. The most well-known mechanisms are biological nitrogen fixation (BNF), where symbiotic or associative bacteria can capture atmospheric nitrogen under microaerobic conditions and through the enzyme nitrogenase, to convert it to forms assimilable by plants. Other mechanisms known are involved in the production of phytostimulatory substances, such as auxin group hormones [5], cytokinins [6], and gibberellins [7].

The constant selection and verification of the effect of plant growth promoting bacteria on species of agronomic interest is necessary for the indication of infective and efficient organisms in the composition of microbial inoculants. Thus, by means of periodic inoculations, it is possible to alter the diversity of the microbial populations interacting with the plants in the rhizosphere, favoring the infection of the roots by efficient and selected microorganisms. With respect to soybean cultivation, for example, in the Brazilian states producing this grain, the reinoculation of the crop induced positive results, compared to the nonreinoculated controls, and in some experiments, increases of up to 23% in yield and up to 25% in the N content of the grains [8]. This contribution favors the economics of mineral fertilizers.

In this chapter, we will discuss the interaction of grasses with soil microorganisms, explain how these microorganisms can benefit the growth and development of grasses, and also elucidate the main forms of interaction between grasses and soil microorganisms.

2. Soil microbial nitrogen (N) transformations

Soil microorganisms, including bacteria, fungi, and protozoa, are responsible for the decomposition of cultural residues such as leaves, stems, and roots, which release significant elements for plant nutrition from organic residues to the mineral phase absorbed by plants. The transformation of nitrogen (N), sulfur (S), and phosphorus (P) is called nutrient cycling. Some simpler composition residues presenting higher concentration of N and P can be easily decomposed. According to Ref. [9], it is related to the chemical composition of the residues, being facilitated by the low ratios of C/N, C/P, lignin/N, polyphenols/N, and (lignin + polyphenols)/N, and difficult because of high levels of lignin and polyphenols.

Plants can absorb N either as Ammonium (NH_4^+) or Nitrate (NO_3^-). In order to achieve that, N must be transformed into a mineral nutrient so that plants can absorb it which depends on the C/N ratio of residue added to the soil. When the C/N ratio is greater than 30/1, the decomposition process is slower than usual, with accumulation of plant residues, as microorganisms cannot easily degrade them. Since the microbial population of the soil lacks nutrients, it competes with plants for N, thus causing a temporary immobilization of N. The C/N ratio greater than 70/1 in grass straws makes the decomposition process more difficult to the soil's microorganisms.

Conversely, when the C/N ratio of plant residues is less than 25/1, N is released [10], thus mineralizing this N present in the soil, which consists in the release of nutrients from the plant residues that plants can absorb as NH_4^+. The legume tissue generally presents a C/N ratio less than 20/1 during the flowering stage. Therefore, after being cut and incorporated into the soil, the legume tissue is a rich source of N to microorganisms which will transform it into a mineral nutrient contributing to the nutrition of grasses and other cultivated plants. As a consequence, part of the mineral N fertilizer can be suppressed in the cultivation of grasses in succession to legumes [11].

Under good drainage conditions, less oxidized forms of N present in the soil, such as ammonium (NH_4^+) and ammonia (NH_3^+), are transformed into more oxidized forms. Nitrifying bacteria of the genera *Nitrosomonas* sp. transform N into volatile nitrite (NO_2–). Fortunately, under the same environmental conditions, *Nitrobacter* sp. transforms volatile nitrite (NO_2–) into nitrate (NO_3–), which is stable and easily absorbed by grasses and other plant families [10].

Under flood conditions, when the supply of O_2 is absent in the soil, some microorganisms carry enzymes capable of consuming the oxygen from the NO_3– present in their respiratory chain as an electron acceptor, transforming it into nitrous oxide (N_2O) [12]. N_2O and other volatile N compounds from microbial activity in poorly drained environments return to the atmosphere as gases. The dinitrogen gas (N_2) can be fixed in the soil through biological N fixation by diazotrophic bacteria. This subject will be discussed individually due to its great importance.

3. Beneficial interaction between grasses and bacteria

Soil bacteria are capable of presenting beneficial effects on cultivated grasses. Several mechanisms make bacteria to promote cultivated grasses, providing significant benefits to the plants, mainly regarding nutritional aspects.

3.1. Beneficial interaction between grasses and nitrogen fixing bacteria

About 78% of the Earth's atmosphere gases are composed of N_2. This gas is neither good nor harmful to mankind. On the other hand, there are in the soil bacteria capable of transforming atmospheric nitrogen (N_2) into nitrogen assimilable by plants (NH_3^+). The enzyme of N-fixing microorganisms that catalyzes the conversion of N into NH_3^+ is named *nitrogenase*. This enzyme is sensitive to oxygen, requiring molecules of iron (Fe), molybdenum (Mo), and vanadium (V) in its structural components [13], besides being an energetically expensive enzyme, requiring two molecules of ATP for each electron [14].

During the 1970s, Döbereiner's findings discovered that bacteria *Azospirillum* and *Herbaspirillum* could endophytically fix N in cultivated grass tissues [15]. There are currently commercial products based on *Azospirillum*, with bacteria selected for maize, wheat [15], and sugarcane [16]. This environmentally friendly process of N-fixing decreases the consumption of mineral N fertilizers, reducing the cost for small farmers, since the demand of industrial N fertilizers with significant consumption of fossil fuels decreases [17].

Besides the endophytic grass fixing, *Azospirillum* and *Herbaspirillum* bacteria can also endophytically fix N in other plants, as in several monocotyledons and dicots such as herbs, shrubs, and trees [18]. When not associated with other leguminous plants [15], these free-living nitrogen-fixing bacteria in the soil are considered optional associative N fixers [19].

Other plant growth promoters include *Azospirillum*, producing phytostimulatory substances, such as the indolyl acetic acid (IAA), gibberellic acid (GA), abscisic acid (ABA), and ethylene [20].

There is also a group of bacteria, named rhizobia, that symbiotically fix the atmospheric N for family *Fabaceae*. Unlike free-living fixers, rhizobia can fix N only when associated within plant root nodules. In the N-fixing symbiosis in leguminous plants belonging to the family *Fabaceae*, the rhizobia receive photo-assimilated carbohydrates and, in exchange, they offer N, which is obtained as N_2 and transformed into NH_3. The *nitrogenase* complex consists of two proteins: Fe-protein and MoFe-protein [3]. Thus, the metabolic exchanges between rhizobia and plants take place in structures called nodules, where the *nitrogenase* is protected from the atmospheric oxygen, due to the presence of leghemoglobin's hemeprotein, presented in high concentrations in active nodules, and fixed to oxygen.

Although they do not directly contribute to the grass nutrition, the symbiotic relationship between legumes and these symbiotic N-fixing bacteria in root nodules promotes the contribution of N to the soil, which will contribute to the nutrition of grasses after the crop cycle of the legume through cultural residues decomposition. There are important reports in the literature on the benefits of grasses grown after legumes interacting with symbiotic N-fixing bacteria [5, 21, 22]. Currently, among the 13 symbiotic N-fixing bacteria, including the genera *Rhizobium*, *Mesorhizobium*, *Bradyrhizobium*, there are the subclasses *α-proteobacteria* and *β-proteobacteria*, with two genera belonging to the *Burkholderiales*, and a genus *Pseudomonas* subclass *γ-proteobacteria* [23].

3.2. The production of phytostimulatory substances by rhizobia

The previous studies including rhizobia as grass growth promoters were driven by empirical findings in which specific rice plants cultivated in succession to clover produced more when compared to a cultivation without succession under the same soil, climate, fertilization, and management conditions, and this increment was not only related to residual N [21].

It could be proven that rhizobia are able to penetrate the grass tissue through fissures and root insertions [24–26]. In the intraradicular environment, as well as in the rhizosphere, rhizobia can produce phytostimulatory substances such as auxins [5, 7], cytokinins [6], and gibberellins [7, 21], which directly favor the yield of cultivated plant species.

It is currently known that rhizobia promote plant growth in interactions with grasses, such as rice [5, 26, 27], barley [28], maize [29], Tanzania grass, and Pensacola [30]. Thus, rhizobia can not only symbiotically fix atmospheric N associated with legumes but they also present a great potential to be exploited as direct promoters of compatible yield increases when inadequately inoculated in succession/rotation cultures systems.

The main phytostimulatory substances produced by rhizobia are the hormones present in auxins [5, 7], cytokinins [6], and gibberellins [7, 21]. The production of abscisic acid [31], lipochitooligosaccharides (LCO) [28, 32], lumichrome [33], aminocyclopropane carboxylic acid deaminase [34], and riboflavones and vitamins [35] produced by rhizobia have also been reported.

Among all auxin syntheses, the indole-3-acetic acid (IAA) is the most studied and the most produced by bacteria [36]. The IAA is mainly presented in the formation of lateral roots and root hair that increase the plant's nutrients absorption [5]. Different metabolic pathways for IAA biosynthesis have already been identified in bacteria [37], being that two main metabolic pathways, indole-3-acetamide (IAM) and indole-3-pyruvate (IpyA), depend on tryptophan [38]. Probably the IAA biosynthesis pathway in rhizobia is the indole-3-acetonitrile (IAN) [39].

Tryptophan has been found in root exudates. Kravchenko et al. [40] quantified the tryptophan exudation by aseptic tomato and radish roots. Tomato seedlings released 2.8–5.3 ng of tryptophan per plant daily, whereas radish seedlings released 190–390 ng of tryptophan per plant per day. In the same study, the authors conducted experiments in soil pots, where they inoculated both cultures with a *Pseudomonas* plant growth-promoting rhizobacteria and observed that radish root mass increased by 36% in inoculated plants, while proven inefficient in tomatoes. The authors state that the beneficial effect of inoculation on radish plants can be explained by the fact that the rhizobacteria produced the plant growth stimulating hormone IAA.

In an experiment conducted in a growth chamber, Silveira [41] studied the effect of inoculation of five *Rhizobium leguminosarum* bv. *trifolii* strains, and their ability to promote rice growth in cultivar IAC103 in nutrient solution. Considering the accumulation of dry mass, plants inoculated with SEMIA235 and SEMIA250 strains were superior compared to the control treatment. The production of IAA by these strains was low, which can be the key to a great plant stimulation. Barazani and Friedman [42] have also reported that deleterious rhizobacteria produced high levels of IAA. However, with plant growth-promoting rhizobacteria, lower yields could be obtained during the same incubation period.

Biswas et al. [5] conducted laboratory and greenhouse studies to test the ability of rhizobia to promote plant growth in two rice cultivars. The studied rhizobia were assessed for the IAA production using the colorimetric method, which was positive for supernatant cultures for all rhizobia tested, ranging from 1.6 to 2.8 $\mu g\ mL^{-1}$. The best responses to inoculations were obtained with *R. leguminosarum* bv. *trifolii* strain E11 and *Rhizobium* sp. strain IRBG74, which presented early stimulation in the plant growth, resulting in an increase in grain and straw yields during the plant's maturity.

Bradyrhizobium japonicum rhizobia isolated from soybean roots, *Azorhizobium caulinodans* isolated from *Sesbania rostrata, Rhizobium* NGR234 isolated from *Lablab purpureus, Sinorhizobium meliloti* isolated from *Medicago sativa, R. leguminosarum* bv. *viceae* Cn6, and *R. leguminosarum* bv. *viceae* strain 30 isolated from *Vicia faba* could infect and colonize sorghum and *Setaria* roots [43]. Considering that this distinct group of rhizobia isolated from different legumes can colonize these two grasses, the authors suggest that the infection of nonlegumes by rhizobia is more likely due to natural conditions than imagined. There was an increase in the growth of inoculated sorghum and *Setaria,* as well as an increase of P in the sorghum. According to the authors, this may have occurred due to the induction of bacteria as phosphate transporters from the plasma membrane of sorghum root cells.

Gibberellins (GAs), phytohormones produced by rhizobia, stimulate stem growth [7, 21], whose effect is mostly observed in grasses, vegetables, and ornamental plants [44]. Important effects of GAs are evident during plant growth, especially on stem elongation, with increased leaf growth and xylem differentiation [45]. At determined GA content, the higher or lower IAA level means the optimal level [46]. Therefore, a certain balance between GA and IAA is essential for the maximum growth rate.

Erum and Bano [7] quantified the production of IAA and GA by rhizobia, using high pressure liquid chromatography (HPLC). The rhizobia, isolated from soil in northern Pakistan, located at 940–3090 m above sea level, produced phytohormones, and the GA production was about 10–30 times higher than the production of IAA. There was a positive and increasing correlation between the GA/IAA ratio produced and this altitude. According to the authors, the IAA and GA concentration gradient may represent a decrease of natural resources, such as radiation intensity, soil moisture, and soil nutrients.

Although cytokinins are produced by rhizobia [6], they have been little studied as it is difficult to detect and quantify them. Cytokines stimulate cell division (cytokinesis), being produced in the plant's root and transported through the xylem to the plant. The levels of auxin and cytokinins are inversely correlated in the plant [47]. Other phytoestimulators produced by rhizobia, the lipochitooligosaccharides (LCO), also known as Nod factors, are responsible for the morphogenetic changes in legume roots during nodulation [48]. They have also stimulated the germination of maize, rice, beet, and cotton, under laboratory, greenhouse, and field conditions [32].

Although the key role of LCO produced by rhizobia in nodule formation is clear, other morphogenetic activities in plants were attributed to LCOs, including the stimulation of genes in the cell division cycle and stimulation of mitotic divisions in protoplasm cultures of legumes and nonlegumes [49].

Miransari and Smith [28] tested the effect of LCO extracted from *B. japonicum* and gibberellin on barley seed germination. In the treatment with 10^{-5} M of gibberellin, there was 18% increase in the seedling germination compared to the control treatment, whereas in the treatment with 10^{-6} M of LCO, the increase represented 44%.

Some rhizobia can lower the level of ethylene excreted by the plants by forming the aminocyclopropane carboxylic (ACC) acid deaminase, an enzyme that breaks ACC, a precursor of ethylene [50]. This enzyme was found in rhizospheric bacteria of the genus *Pseudomonas,*

Alcaligenes, *Rhodococcus*, and *Rhizobium* [34, 51]. In addition, some bacteria of *Rhizobium japonicum* (*B japonicum*) species synthesize phytotoxic antibiotics, aminoethoxyvinylglycine, and rhizobitoxin, which inhibit the formation of ethylene in plants [52]. The ethylene is a plant growth inhibitor, therefore bacteria that regulate its production can indirectly stimulate the plants and may be associated with the cell development, cell extension and the postponement of the fall of leaves and fruits [52].

S. meliloti can produce lumichrome [33]. The application of lumichrome in nanomolecular concentrations promoted the growth of legumes and grasses [43]. According to Ref. [48], the lumichrome stimulated the photosynthetic index of maize on the first and second day after application. Gouws [53] reported an increase in the root biomass of *Lotus japonicus* and tomato when treated with lumichrome. According to the author, the treatment with lumichrome caused complex changes in the gene expression of *L. japonicus* and tomato, being mainly affected the genes associated with the transcription regulation and ribonucleic acid (RNA) signaling, synthesis, degradation, proteins modification, and plant stress responses. The mechanism by which lumichrome promotes the plants growth still needs to be clarified.

3.3. Other phytostimulating-producing bacteria

Other microorganisms, such as *Azospirillum* spp. [54, 55], *Acetobacter diazotrophicus* [56], *Herbaspirillum seropedicae* [56], *Klebsiella pneumoniae* [55], *Pseudomonas syringae* [57], and *Paenibacillus polymyxa* [58], also produce phytostimulatory substances and are also related to the stimulation of grasses and other nonleguminous species.

3.4. Siderophore-producing bacteria

Siderophores are iron-chelating compounds, nutrients that limit the microbial population growth in the soil's environment. Fe must be present as Fe^{2+}, and many microorganisms such as bacteria and fungi have developed mechanisms to chelate Fe^{3+} through the production of siderophores before being transformed into Fe^{2+}.

Siderophores are Fe sequestrants of high affinity and low molecular weight. Among the siderophores known, pyoverdine and enterobactin are secreted by microorganisms in response to the low availability of Fe_3^+ in solution [59]. The siderophores are iron-chelating compounds, entering the cell without reducing Fe^{3+} [60]. Thus, siderophores can capture Fe_3^+ produced under ion-deficiency conditions by fungi and bacteria in order to incorporate this mineral into the cell metabolism [61].

Because of their great ability to compete for the cell metabolism, microorganisms producing siderophores are capable of suppressing the growth and development of pathogenic microorganisms that inhabit the rhizosphere, thus indirectly contributing to the health of cultivated species of plants, such as grasses. Some rhizobacteria of the *genera Pseudomonas* can produce iron-chelating compounds, present in low concentrations in the rhizosphere, and thus suppressing the presence of pathogens near the roots [62].

3.5. Phosphate solubilizing bacteria

Together with N and K, P is one of the required macronutrients for the cultivation of grasses, whose content concentration is always lower than N and K. However, it is commonly necessary to use a great amount of phosphate fertilizers in agricultural crops, because in spite of the soils contain a large amount of P their availability to the plants is very little as P tends to form very low solubility compounds in the soil [63].

Phosphorus is an essential element to grasses, since it is necessary and irreplaceable for the composition of ribonucleic acids (RNA) and deoxyribonucleic acids (DNA), responsible for the transmission of the genetic code to the plants, protein production, and other essential compounds for the plant structure and seedling production. Grasses absorb soil P as $H_2PO_4^-$ and HPO_4^{2-}, just like other plant species; thus, insoluble phosphates like tricalcium phosphate ($Ca_3(PO_4)_2$) make this nutrient unavailable to the plant. Some of the soil bacteria are important in the process of dissolving these insoluble solutions, facilitating the access to this essential nutrient.

Inorganic phosphate-solubilizing microorganisms excrete inorganic acids and protons associated to these acids, which directly dissolve the insoluble phosphate, or chelate the cations with the phosphate anion [64]. Among phosphate-solubilizing bacteria, *Burkholderia* [65], *Bacillus*, and *Penicillium* strains [66] have been reported.

4. Mycorrhizae

Mycorrhizal fungi are associated with roots of plants and play an important role in the soil phosphorus cycling as extensions of the root system, increasing the absorbing area of the root and the absorption rate of phosphorus. The mycorrhizal association does not substitute phosphate fertilization, but efficiently increases the use of phosphorus or an added compound through fertilization [67]. Grasses such as maize, sorghum, wheat, rice, and cultivated forage grasses may have their roots naturally colonized by mycorrhizal fungi [68].

5. Fungi of the genus *Trichoderma* and the biological control of diseases

Fungi of the genus *Trichoderma* are biological control agents that act against phytopathogenic edaphic fungi, however, colonizing plant roots to stimulate plant growth and protect them against infections. Root colonization often increases root development, crop productivity, resistance to abiotic stresses, and improved nutrient use [69].

Fungi present different mechanisms to controlling and suppressing the soil's phytopathogens, such as mycoparasitism, antibiosis and antagonism [70]. In addition, *Trichoderma* strains are active in the production of fungal cell walls, enzymatic degradation, including pectinases, cellulases, and chitinases, involved in the biological control [71].

Compared to the chemical control, the use of biological substances to control soil diseases is beneficial, since it does not induce resistance from the target organism, effectively controlling it in a long term. There are currently commercial products based on genus *Trichoderma* strains, properly registered in the Brazilian Ministry of Agriculture, Livestock, and Food Supply (MAPA) that are indicated for the controlling of diseases caused by phytopathogenic agents, such as *Rhizoctonia*, *Fusarium*, and *Sclerotinia* [72]. Harman [4] and Machado et al. have described the benefits of *Trichoderma* inoculation on grass yield, observing an increase in maize and black oat yields, respectively.

6. Final considerations

Soil microorganisms are able to influence the establishment and yield of grasses by means of several mechanisms. The nutrients cycling by soil microorganisms, the biological nitrogen fixation by associative bacteria, phytohormones production by soil bacteria, and the acquisition of phosphorus by mycorrhizal fungal hyphae networks are just some examples of direct mechanisms of beneficial interaction between soil microorganisms and cultivated grasses. As examples of indirect mechanisms, we can mention the suppression of pathogens by mechanisms of predation or competition, as we also discussed. Given this wide range of mechanisms presented by microorganisms for the benefit of cultivated grasses and consequently of the human benefit, it is imperative that these mechanisms are well studied to be inserted in systems of conservationist agriculture, which must obtain the maximum agronomic yield of the crops, allied to the rational use of natural resources.

Author details

Rafael Goulart Machado

Address all correspondence to: rgoulartmachado@gmail.com

1 Emater-RS/ASCAR, Brazil

2 College of Agronomy, Institute of Educational Development from Passo Fundo, Brazil

References

[1] Gans J, Wolinsky M, Dunbar J. Computational improvements reveal great bacterial diversity and high metal toxicity in soil. Science. 2005;**309**:1387-1390. DOI: 10.1126/science.1112665

[2] Bashan Y, Holguin G. *Azospirillum*-plant relationships: Environmental and physiological advances. Canadian Journal of Microbiology. 1997;**43**:103-121 DOI: 10.1139/m97-015

[3] Taiz L, Zieger E, editors. Fisiologia Vegetal. 3rd ed. Porto Alegre: Artemed; 2004. p. 719

[4] Harman GE. Myth and dogmas of biocontrol changes in perceptions derived from research on *Trichoderma harzianum* T-22. Plant Disease. 2000;**84**:377-393. DOI: 10.1094/PDIS.2000.84.4.377

[5] Biswas JC, Ladha JK, Dazzo FB, Yanni YG, Rolfe BG. Rhizobial inoculation influences seedling vigor and yield of rice. Agronomy Journal. 2000;**92**:880-886. DOI: 10.2134/agronj2000.925880x

[6] Persello-Cartieaux F, Nussaume L, Robaglia C. Tales from the underground: Molecular plant-rhizobacteria interactions. Plant, Cell and Environment. 2003;**26**:189-199. DOI: 10.1046/j.1365-3040.2003.00956.x

[7] Erum E, Bano A. Variation in phytohormone production in *Rhizobium* strains at different altitudes of north areas of Pakistan. International Journal of Agriculture and Biology. 2008;**10**:536-540

[8] Hungria M, Campos RJ, Mendes IC. Fixação biológica do nitrogênio na cultura da soja. Londrina: EMBRAPA; 2001. p. 48

[9] Carvalho AM, Bustamante MMC, Alcântara FA, Resck IS, Lemos SS. Characterization by solid-state CPMAS ^{13}C NMR spectroscopy of decomposing plant residues in conventional and no-tillage systems in Central Brazil. Soil & Tillage Research. 2009;**102**:144-150. DOI: 10.1016/j.still.2008.08.006

[10] Camargo FAO, Sá ELS. Nitrogênio e adubos nitrogenados. In: Bissani CA, Gianello C, Tedesco MJ, Camargo FAO, editors. Fertilidade dos solos e manejo da adubação de culturas. Porto Alegre: Genesis; 2004. pp. 93-116

[11] Sociedade Brasileira de Ciência do Solo. Comissão de Química e Fertilidade do Solo. Manual de adubação e calagem para os Estados do Rio Grande do Sul e de Santa Catarina. 10th ed. Porto Alegre: SBCS, Núcleo Regional Sul; 2004. p. 400

[12] Dalal RC, Wang W, Robertson GP, Parton WJ. Nitrous oxide emission from Australian agricultural lands and mitigation options: A review. Australian Journal of Soil Research. 2003;**41**:165-195. DOI: 10.1071/SR02064

[13] Reis VM, Teixeira KRS. Fixação biológica de nitrogênio - estado da arte. In: Aquino AM, Assis RL, editors. Processos biológicos no sistema solo-planta: Ferramentas para uma agricultura sustentável. Brasília: EMBRAPA; 2006. pp. 151-180

[14] Howard JB, Rees DC. Structural basis of biological nitrogen fixation. Chemical Reviews. 1996;**96**:2965-2982

[15] Hungria M. Inoculação com *Azospirillum brasilense*: inovação em rendimento a baixo custo. Londrina: EMBRAPA; 2011. p. 38

[16] Reis VM, Baldani JI, Urquiaga S. Recomendação de uma mistura de estirpes de cinco bactérias fixadoras de nitrogênio para inoculação de cana de açúcar: *Gluconacetobacter diazotrophicus* (BR 11281), *Herbaspirillum rubrisubalbicans* (BR 11504), *Azospirillum amazonense* (BR 11145) e *Burkholderia tropica* (BR 11366). Seropédica: EMBRAPA; 2009. p. 4

[17] Howieson JG, McInnes A. The legume rhizobia symbiosis. Does it vary for the tropics relative to the Mediterranean basin?. In: Proceedings of the XIX international grasslands congress. São Pedro: Brazilian Society of Animal Husbandry; 2001. pp. 585-590

[18] Lange A, Moreira FMS. Detecção de *Azospirillum amazonense* em raízes e rizosfera de Orchidaceae e de outras famílias vegetais. Revista Brasileira de Ciência do Solo. 2002;**26**:529-533. DOI: 10.1590/S0100-06832002000200027

[19] Baldani J, Caruso L, Baldani VLD, Goi SR, Döbereiner J. Recent advances in BNF with non-legume plants. Soil Biology & Biochemistry. 1997;**29**:911-922. DOI: 10.1016/S0038-0717(96)00218-0

[20] Perrig D, Boiero ML, Masciarelli OA, Penna C, Ruiz OA, Cassán FD, Luna MV. Plant-growth-promoting compounds produced by two agronomically important strains of *Azospirillum brasilense*, and implications for inoculant formulation. Applied Microbiology and Biotechnology. 2007;**75**:1143-1150. DOI: 10.1007/s00253-007-0909-9

[21] Yanni, YG, Rizk RY, El-Fattah FKA, Squartini A, Corich V, Giacomini A, de Bruijn F, Rademaker J, Maya-Flores J, Ostrom P, Vega-Hernandez M, Hollingsworth RI, Martinez-Molina E, Mateos P, Velazquez E, Wopereis J, Triplett E, Umali-Garcia M, Anarna JA, Rolfe BG, Ladha JK, Hill J, Mujoo R, Ng PK, Dazzo FB. The beneficial plant growth-promoting association of *Rhizobium leguminosarum* bv. *trifolii* with rice roots. Australian Journal of Plant Physiology. 2001;**28**:845-870. DOI: 10.1071/PP01069

[22] Hahn L, Sá ELS, Osório Filho BD, Machado RG, Damasceno RG, Giongo A. Rhizobial inoculation alone or coinoculated with *Azospirillum brasilense*, promotes growth of wetland rice. Revista Brasileira de Ciência do Solo. 2016;**40**:1-15. DOI: 10.1590/1806965 7rbcs20160006

[23] Weir BS. The Current Taxonomy of Rhizobia. NZ Rhizobia website. 2016. Available from: https://www.rhizobia.co.nz/taxonomy/rhizobia. (Accessed: 19-02-2017])

[24] Reddy PM, Ladha JK, So RB, Hernandez RJ, Ramos MC, Angeles OR, Dazzo FB, de Bruijn FJ. Rhizobial communication with rice roots: Induction of phenotypic changes, mode of invasion and extent of colonization. Plant and Soil. 1997;**194**:81-98. DOI: 10.1023/A:1004243915997

[25] Webster G, Gough C, Vasse J, Batchelor CA, O'Callaghan KJ, Kothari SL, Davey MR, Dénarié J, Cocking EC. Interactions of rhizobia with rice and wheat. Plant and Soil. 1997;**194**:115-122. DOI: 0.1023/A:1004283819084

[26] Yanni YG, Rizk RY, Corich V, Squartini A, Ninke K, Hollingsworth SP, Orgambide G, de Bruijn F, Stoltzfus J, Buckley D, Schmidt TM, Mateos PF, Ladha JK, Dazzo FB. Natural endophytic association between *Rhizobium leguminosarum* bv. trifolii and rice roots and assessments of its potential to promote rice growth. Plant and Soil. 1997;**194**:99-114. DOI: 10.1023/A:1004269902246

[27] Osório Filho BD, Gano KA, Binz A, Lima RF, Aguilar LM, Ramirez A, Caballero-Mellado J, Sá ELS, Giongo A. Rhizobia enhance growth in rice plants under flooding conditions. American-Eurasian Journal Agricultural and Environmental Science. 2014;**14**:707-718. DOI: 10.5829/idosi.aejaes.2014.14.08.12377

[28] Miransari M, Smith, D. Rhizobial lipo-chitooligosaccharides and gibberellins enhance barley (*Hordeum vulgare* L.) seed germination. Biotechnology. 2009;**8**:270-275. DOI: 10.3923/biotech.2009.270.275

[29] Hahn L, Sá ELS, Machado RG, Silva WR, Oldra S, Damasceno RG, Schönhofen A. Growth promotion in maize with diazotrophic bacteria in succession with ryegrass and white clover. American-Eurasian Journal of Agriculture and Environmental Science. 2014;**14**:11-16. DOI: 10.5829/idosi.aejaes.2014.14.01.11893

[30] Machado RG, Sá ELS, Bruxel M, Giongo A, Santos NS, Nunes AS. Indoleacetic acid producing *Rhizobia* promote growth of Tanzania grass (*Panicum maximum*) and Pensacola grass (*Paspalum saurae*). International Journal of Agriculture and Biology. 2013;**15**:827-834

[31] Dangar TK, Basu PS. Abscisic acid production in culture by some *Rhizobium* spp. of leguminous trees and pulses. Folia Microbiologica. 1991;**36**:527-532. DOI: 10.1007/BF02884031

[32] Prithiviiraj B, Zhou X, Souleimanov A, Khan WM, Smith DL. A host-specific bacteria-to-plant signal molecule (Nod factor) enhances germination and early growth of diverse crop plants. Planta. 2003;**216**:437-445. DOI: 10.1007/s00425-002-0928-9

[33] Volpin H, Phillips DA. Respiratory elicitors from *Rhizobium meliloti* affect intact alfalfa roots. Plant Physiology. 1998;**116**:777-783

[34] Ma W, Guinel F, Glick B. *Rhizobium leguminosarum* biovar viciae 1-aminocyclopropane-1-carboxylate deaminase promotes nodulation of pea plants. Applied and Environmental Microbiology. 2003;**69**:4396-4402. DOI: 10.1128/AEM.69.8.4396-4402.2003

[35] Dakora FD. Defining new roles for plant and rhizobial molecules in sole and mixed plant cultures involving symbiotic legumes. New Phytologist. 2003;**158**:39-49. DOI: 10.1046/j.1469-8137.2003.00725.x

[36] Radwan TSD, Mohamed ZK, Reis VM. Aeração e adição de sais na produção de ácido indol acético por bactérias diazotróficas. Pesquisa Agropecuária Brasileira. 2005;**40**:997-1004. DOI: 10.1590/S0100-204X2005001000008

[37] Spaepen S, Vanderleyden J, Remans R. Indole-3-acetic acid in microbial and microorganism-plant signaling. FEMS Microbiology Reviews. 2007;**31**:425-448. DOI: 10.1111/j.1574-6976.2007.00072.x

[38] Lambrecht M, Okon Y, Vande Broek A, Vanderleyden J. Indole-3-acetic acid: A reciprocal signalling molecule in bacteria-plant interactions. Trends in Microbiology. 2000;**8**: 298-300

[39] Osorio Filho BD. Rizóbios eficientes em *Lotus* em condições de estresse hídrico e promotores de crescimento de arroz irrigado. Porto Alegre: Soil Science Graduate Program, UFRGS; 2009. p. 113

[40] Kravchenko LV, Azarova TS, Makarova NM, Tikhonovich IA. The effect of tryptophan present in plant root exudates on phytostimulating activity of rhizobacteria. MAIK Nauka/Interperiodica – Microbiology. 2004;**73**:156-158. DOI: 10.1023/B:MICI.0000023982.76684.9d

[41] Silveira EL. Inoculações de bactérias promotoras de crescimento no cultivo de arroz em solução nutritiva. Jaboticabal: Faculty of Agrarian and Veterinary Sciences, State University of São Paulo; 2008. p. 99

[42] Barazani O, Friedman J. Is IAA the major root growth factor secreted from plant-growth-mediating bacteria?. Journal of Chemical Ecology. 1999;**25**:2397-2406. DOI: 10.1023/A:1020890311499

[43] Matiru VN, Dakora FD. Potencial use of rhizobial bacteria as promoters of plant growth for increased yield in landraces of African cereal crops. African Journal of Biotechnology. 2004;**3**:1-7

[44] Weaver RJ. Plant Growth Substances in Agriculture. San Francisco: W.H. Freeman and Company; 1972. p. 594

[45] Metivier JR. Giberelinas. In: Ferri MG, editor. Fisiologia Vegetal. 2nd ed. São Paulo: EDUSP; 1986. pp. 129-161

[46] AndersAnderson IC. Plant characteristic that affect yeld. In: Proceedings of the Hybrid corn industry research conference. 22nd ed. Washington: HCIRC`67; 1967. p. 71-73

[47] Eklöf S, Astot C, Sitbon F, Moritz T, Olsson O, Sandberg G. Transgenic tobacco plants co-expressing *Agrobacterium iaa* and *ipt* genes have wild-type hormone levels but display both auxin- and cytokinin-overproducing phenotypes. Plant Journal. 2000;**23**:279-284. DOI: 10.1046/j.1365-313x.2000.00762.x

[48] Khan W, Prithiviraj B, Smith DL. Nod factor [Nod Bj V (C18:1, MeFuc)] and lumichrome enhance photosynthesis and growth of corn and soybean. Journal of Plant Physiology. 2008;**185**:1342-1351. DOI: 10.1016/j.jplph.2007.11.001

[49] Souleimanov A, Prithiviraj B, Smith DL. The major Nod factor of *Bradyrhizobium japonicum* promotes early growth of soybean and corn. Journal of Experimental Botany. 2002;**53**:1929-1934. DOI: 10.1093/jxb/erf034

[50] Penrose D, Glick B. Determination of 1-aminocycopropane-1-carboxylic acid (ACC) to assess the effects of ACC deaminase-containing bacteria on roots of canola seedlings. Canadian Journal of Microbiology. 2001;**47**:77-80. DOI: 10.1139/w00-128

[51] Belimov AA, Safronova VI, Sergeyeva TA, Egorova TN, Matveyeva VA, Tsyganov VE, Borisov AY, Tikhonovich IA, Kluge C, Preisfeld A, Dietz KJ, Stepanok VV. Characterization of plant growth promoting rhizobacteria isolated from polluted soils and containing 1-aminocyclopropane-1-carboxylate deaminase. Canadian Journal of Microbiology. 2001;**47**:642-652. DOI: 10.1139/cjm-47-7-642

[52] Tsavkelova EA, Klimova SY, Cherdyntseva TA, Netrusov AI. Hormones and hormone-like substances of microorganisms: A review. Applied Biochemistry and Microbiology. 2006;**42**:229-235. DOI: 10.1134/S000368380603001X

[53] Gouws LM. The Molecular Analysis of the Effects of Lumichrome as a Plant Growth Promoting Substance. Stellenbosch: Stellenbosch University; 2009. p. 161

[54] Van der Broek A, Vanderleyden J. Review: Genetics of the *Azzospirillum*-plant root association. Critical Reviews in Plant Sciences. 1995;**14**:445-446. DOI: 10.1080/07352689509701932

[55] El-Khawas H, Adachi K. Identification and quantification of auxins in culture media of *Azospirillum* and *Klebsiella* and their effect on rice roots. Biology and Fertility of Soils. 1999;**28**:377-381. DOI: 10.1007/s003740050507

[56] Bastián F, Cohen A, Piccoli P, Luana V, Bottini R, Baraldi R, Bottini R. Production of índole-3-acetic acid and gibberellins A1 and A3 by *Acetobacter diazotrophicus* and *Herbaspirillum seropedicae* in chemically defined culture media. Journal of Plant Growth Regulation. 1998;**24**:7-11. DOI: 10.1023/A:1005964031159

[57] Weingart H, Völksch B. Ethylene production by *Pseudomonas syringae* Pathovars in vitro and in planta. Applied and Environmental Microbiology. 1997;**63**:156-161. DOI: 10.1007/978-94-011-5472-7_59

[58] Lebuhn M, Heulin T, Hartmann A. Production of auxin and other indolic and phenolic compounds by *Paenibacillus polymyxa* strains isolated from different proximity to plant roots. FEMS Microbiology Ecology. 1997;**22**:325-334. DOI: 10.1016/S0168-6496(97)00007-X

[59] Oliveira MS. Seleção de rizobactérias autóctones para a promoção de crescimento de feijoeiro comum (*Phaseolus vulgaris* L.). Porto Velho: Union of Higher Education from Rondônia; 2010. p. 45

[60] Santos DMM. Disciplina de Fisiologia Vegetal. Jaboticabal: UNESP; 2004. p. 8

[61] Silva DESG, Oliveira LG. Avaliação da produção de sideróforos em actinomicetos. In: Proceedings of the XIX Internal Congress of Scientific Initiation. Campinas: UNICAMP; 2012. pp. 237-237

[62] Vieira Júnior JR, Fernandes CF, Antúnes Júnior H, Silva MS, Silva DSG, Silva UO. Rizobactérias como agentes de controle biológico e promotores de crescimento de plantas (Documentos 155). Porto Velho: Embrapa; 2013. p. 15

[63] Bissani CA, Gianello C, Camargo FAO, Tedesco MJ. Fertilidade dos solos e manejo da adubação de culturas. 2nd ed. Porto Alegre: Editora Metrópole; 2008. p. 344

[64] Richardson AE. Soil microorganisms and phosphorus availability: In: Pankhurst CE, Doube BM, Gupta VVSR, Grace PR, editors. Soil Biota Management in Sustainable Farming Systems. Melbourne: CSIRO; 1994. p. 50-62

[65] Alves JB. Seleção de rizóbios para trevo branco. Porto Alegre: Soil Science Graduate Program, UFRGS; 2005. p. 78

[66] Mendes IC, Reis Junior FB. Microorganismos e disponibilidade de fósforo (P) nos solos: Uma análise crítica (Documentos 85). Planaltina: Embrapa; 2003. p. 26

[67] Miranda JCC, Miranda LN. Micorriza arbuscular. In: Vargas MAT, Hungria M, editors. Biologia dos solos dos cerrados. Planaltina: Embrapa; 1997. pp. 69-123

[68] Zambolim L, Siqueira JO. Importância e potencial das associações micorrízicas para a agricultura (Documentos 26). Belo Horizonte: EPAMIG; 1985. p. 36

[69] Benítez T, Rincón AM, Limón MC, Codón AC. Biocontrol mechanisms of Trichoderma strains. International Microbiology. 2004;**7**:249-260

[70] Machado DFM, Parzianello FR, Silva ACF, Antoniolli ZI. Trichoderma no Brasil: o fungo e o bioagente. Revista de Ciências Agrárias. 2012;**35**:274-288

[71] Barbosa MAG, Rehn KG, Menezes M, Mariano RLR. Antagonism of *Trichoderma* species on *Cladosporium herbarum* and their enzymatic characterization. Brazilian Journal of Microbiology. 2001;**32**:98-104. DOI: 10.1590/S1517-83822001000200005

[72] Ministry of Agriculture Livestock and Food Supply (MAPA). Agrofit, Sistema de agrotóxicos fitossanitários. 2017. Available from: http://agrofit.agricultura.gov.br/agrofit_cons/!ap_produto_form_detalhe_cons?p_id_produto_formulado_tecnico=8300&p_tipo_janela=NEW [Accessed: 27-01-2017]

Chapter 3

Importance of Grass Carp (*Ctenopharyngodon idella*) for Controlling of Aquatic Vegetation

Yusuf Bozkurt, İlker Yavas, Aziz Gül,
Beytullah Ahmet Balcı and Nurdan Coskun Çetin

Additional information is available at the end of the chapter

http://dx.doi.org/10.5772/intechopen.69192

Abstract

Aquatic plants are beneficial and a necessary part of lakes and reservoirs. Also, some kind of plants are the main food source for aquatic animals. Plants are able to stabilize sediments, improve water clarity and add diversity to the shallow areas of lakes. On the other hand, overgrown plants can become a nuisance by hindering human uses of water and threaten the structure and function of diverse native aquatic ecosystems. This chapter aims to make analysis of using of grass carp to control aquatic vegetation. In this concept, origin and distribution, biological features, reproduction, feeding behaviour and effects of grass carp on aquatic plants, water body and sediments are also discussed.

Keywords: aquatic plants, grass carp, biological control, ecology, habitat

1. Introduction

Aquatic plants are important elements in water reservoirs. They play a fundamental role in energy and carbon storage at the bases of food pyramids. In addition, they act as protection and reproduction refuges for many organisms, and its submerged parts allow the development of periphyton communities [1].

Unfortunately, as beneficial as they are, aquatic plants can easily overpopulate and become a nuisance to the landowner. Plants can also harm the fishing potential of the water body. An excess of decaying plants can lower the amount of oxygen in the water that can be harmful for the aquatic species. In some waters, abundance of plants overprotects fish and other prey species allowing them to overpopulate.

Controlling and eliminating aquatic vegetation from ponds are often confusing and frustrating tasks. The selection of a vegetation control program depends on local conditions of the pond. For this aim, there are three approaches including mechanical, chemical and biological control. First one is the mechanical control which involves physical removal of the vegetation and is often more difficult in water than on land. Second, chemical vegetation control is often unsuccessful, and retreatment may be needed. Also, chemical vegetation control can become expensive, and the selection of a chemical depends on the plant species involved. In addition, chemical vegetation control is short lived due to most of aquatic herbicides that do not persist more than a few months.

For these reasons, the ideal aquatic plant management tool should provide cost effective control with long-term impact, a high level of selectivity and if possible have minimal or no negative side effects. Another alternative control method to mechanical or chemical vegetation control is biological control which involves using of fishes to control the aquatic vegetation. Biological control has many advantages over the other vegetation control means. For instance, it takes much less human work effort than most of mechanical control means and does not require using expensive and hazardous aquatic herbicides. In addition, using fish species provides longer term control than other control mechanisms due to fishes that usually have a life-span of several years.

Fish used for aquatic vegetation control include several species of tilapia (*Tilapia* spp.), silver carp (*Hypophthalmichthys molitrix*) and the grass carp (*Ctenopharyngodon idella*). Of these fish, only the grass carp is able to consume large quantities of aquatic macrophytes [2]. Under suitable conditions, adult grass carp can consume more than its own weight of plant material on a daily basis [3].

From this point of view, controlling aquatic vegetation with grass carp is one of the available options for pond owners with aquatic plant problems. In many situations, using grass carp is an economical, long-lasting and effective option.

2. The grass carp

Grass carp (*C. idella*), also known as the white Amur, is one of the most important farmed freshwater fish species with an annual global production of 5,537,794 tons in the year of 2014 [4]. The grass carp is one of the largest members of the family *Cyprinidae* and is the only member of the genus *Ctenopharyngodon* [5, 6]. It shouldn't be confused with other carp species such as silver carp (*Hypophthalmichthys molitrix*), bighead carp (*Hypophthalmichthys nobilis*) or mud carp (*Cirrhinus molitorella*) These carp species are not good biological control agents for aquatic vegetation because they feed on different components of the pond ecosystem.

Grass carp is native to southeastern Russia and northwestern China. This herbivorous species has been deliberately introduced into many countries for vegetation control purposes. In addition, the grass carp is an integral part of fish culture and forms an important source of protein for human consumption.

2.1. Distribution of grass carp

Grass carp is a sub-tropical-to-temperate species and is native to large rivers and lakes in eastern Asia. Its native range extends from southern Russia southward to northern Vietnam and in large rivers like the Amur (border of China and Russia), Yang Tze (northern China), Yellow River (central China), and the Min River (crosses the border from Vietnam into China) [5].

In addition, grass carp have been introduced to many countries around the world including Taiwan, Israel, Japan, the Philippines, the United States, Mexico, India, Malaysia, the Netherlands, Switzerland, Czechia, Slovakia, Denmark, Sweden, Romania, Poland, Italy, West Germany, France, the United Kingdom, Argentina, Venezuela, Fiji, New Zealand, Australia and South Africa [5].

Grass carp are considered uncommon in their Amur basin native range, relative to other species of Asian carps. There is a broad range of climatic conditions within the native range of the grass carp. The mean annual air temperatures range from 25°C (in the southernmost part of the hemisphere) to −60°C (in the northernmost part of the hemisphere) [5].

2.2. Biological features of grass carp

Grass carp is characterized with a wide and scale-less head, sub-terminal or terminal mouth with simple lips which do not include barbels, protracted upper jaw and a very short snout [7–9].

The body is slender and rather compressed with a rounded belly and slightly decurved lateral line [9]. Dorsal fin origin is above or just in front of the pelvic fin origin and the dorsal and anal fins do not have spines [5, 10]. Cycloid scales are dark-edged with a black spot at the base, and the gill rakers are short, lanceolate and widely set [7, 9]. Pharyngeal teeth are bi-serial and are 2.5–4.2, 2.4–4.2, 2.4–5.2 or 1.4–5.2 [5]. Diploid chromosome number is n = 48 and biochemical analysis of five tissues revealed an estimated 49 loci [9]. The colour of adult grass carp is dark grey on the dorsal surface with lighter sides (white to yellow) that have a slightly golden shine. Fins are clear to grey-brown colour [7].

2.3. Reproduction of grass carp

Mature grass carp require approximately 1500 to 2000 days within a year for gonadal development and maturation [11]. Maturity occurs at earlier ages and smaller sizes in tropical climates [5] which is between the ages of 1 to 8 years in the introduced and cultured grass carp populations. Grass carp males generally mature 1 year earlier than females at 50–86 cm in length [5].

In grass carp, the external sexual dimorphism appears in adults at the onset of maturity with the appearance of tubercles on the dorsal and medial surfaces of the pectoral fins in males. Temporary tubercles may develop in females, but they are not as highly developed as in the males. Females exhibit soft, bulging abdomens and swollen, pinkish vents at onset of maturity [5].

On the other hand in some temperate regions, in spite of grass carp maturing at the same time as in their native distribution, their gonads do not mature. This is possibly related to a lack of nutritional, photoperiod and water temperature requirements for grass carp [12]. A well-marked and limited spawning season occurs in temperate latitudes. On the other hand, in tropical areas, the breeding season expands and becomes less distinct, and as a result of this, multiple spawning can occur in a year [5].

In their native areas, grass carp begin migration to spawning areas when water temperatures reach 15–17°C [6]. Water temperature and its level play key roles for inducing spawning, and it varies with latitude. Water temperature required for the stimulation of sexual maturation and spawning ranges between 20 and 30°C. Optimum spawning temperature is generally thought to be between 20 and 22°C. In addition, increases in water level exceeding 122 cm within a 12-hour period are required for spawning [6]. If water levels do not rise during the spawning season, females with small reserves of body fat will either release no eggs or release only a portion. Non-released eggs are subsequently absorbed in the body [13].

Grass carp spawn in the rivers and canals during high water. Spawning usually takes place in spring and summer in the upper part of the water column over rapids or sand bars [5]. Preferred spawning habitat is found in turbulent water of the junction of rivers or below dams [14, 15]. Grass carp prefer to spawn in water currents ranging from 0.6 and 1.5 m/sec, but spawns generally occur in currents as low as 0.2 m/sec or even in ponds where the current is absent [15].

Fecundity is directly proportional to length, weight and age of the females and ranges from 0.001- to 2-million eggs but generally averages to 0.5 million for a 5-kg broodstock [5, 6]. Grass carp eggs are 2.0–2.5 mm in diameter when released but quickly swell to a diameter of 5–6 mm as water is absorbed [6]. The eggs are semi-buoyant and nonadhesive, requiring well-oxygenated water and a current to keep them suspended until hatching [6, 15, 16]. Eggs may travel along the downstream, that's about 50–180 km [14].

2.4. Feeding behaviour of grass carp

Grass carp feed almost exclusively on aquatic plants. They can eat 2–3 times their weight each day and may gain 2–4 kg in a single year. The larger they get, the more plant material they consume. Cultured grass carp may reach up to 1 kg in the first year and grow approximately 2–3 kg/year in temperate areas and 4.5 kg/year in tropical areas [5].

Grass carp prefer soft and low fibre aquatic vegetation such as duckweed and various underwater plants. If the more desired plant species aren't available, they feed on plants above of the water surface. Grass carp even have been observed to feed on terrestrial plants that are hanging over the water. Triploid and diploid grass carp seem to consume similar quantities of aquatic plants and to have similar feeding habits and prefer succulent young plants. Because of its strong preference for aquatic vegetation, the grass carp is being widely used to control aquatic vegetation in lakes and ponds [7].

The five most-preferred species in order of preference are hydrilla, musk grass, pondweeds (*Potamogeton* spp.), southern naiad (*Najas guadalupensis*) and Brazilian elodea (*Egeria densa*) [17]. Grass carp are not a good control method for filamentous algae, Eurasian milfoil (*Myriophyllum spicatum*), spatterdock (*Nuphar advena*), fragrant water lily (*Nymphaea odorata*), sedge (*Cladium* spp.), cattail (*Typha* spp.) or other large aquatic plants [18]. Factors such as age, size, temperature, availability of plant species, size of waterbody and stocking density (in pond cultures) may influence grass carp feeding strategies [9].

While active feeding begins at 7–8°C, intensive feeding occurs only when water temperature is at least 20°C [16]. Three or four days after hatching, larval grass carp begin feeding on rotifers and protozoans, moving up to larger cladocerans at 11–15 days after hatch [9, 14]. By 2 weeks after hatching, grass carp feed on larger prey such as daphnia and insect larvae [9, 14]. After 3 weeks, the occurrence of plants in the diet increases with the appearance of filamentous algae and macrophytes. Macrophyte feeding begins from 1 to 1.5 months after hatching [9]. However, juveniles consume other items including chironomids, cladocerans, copepods, insects and their aquatic larvae, crustaceans and small fishes [6].

Studies indicate that grass carp lose weight when kept in unvegetated ponds with sufficient animal food sources [19]. When the supply of macrophytes is low, adult grass carps are able to utilize other food sources including benthos, zooplankton, water beetles and crayfishes [16]. Lopinot [20] indicated that grass carp feeds on almost anything when vegetated food is scarce including small fishes, worms and insects, but in pond culture, they seem to prefer pelleted food to vegetation.

3. The method

Intensive use of chemical fertilizers in agriculture and also human and industrial pollution causes eutrophication. This situation causes growing of plants quickly and as a result of this, plant control cannot be solved mechanically or chemically. The most obvious solution in these cases is the introduction of grass carp to these waters covered with plants.

Some several thousand hectares of large ponds covered with overgrown macrophytic vegetation can be cleaned by introducing of grass carp. Grass carp is one of the optimal species for controlling of aquatic plants in water reservoirs. At this point, several parameters such as stocking density of grass carp, plant and plankton composition, water quality, and also the structure of the benthos should be noted.

3.1. Stocking of grass carp for controlling of aquatic vegetation

The grass carp number required to control aquatic plants varies depending on the degree of plant infestations, plant types, pond sizes and the size of fishes stocked. A number of different methodologies have been used to determine the suitable number of grass carp to stock. The most precise method is to determine the weight of aquatic vegetation in the pond and knowing the consumption rates of the fish.

In spite of investigation of different stocking rates, there is no guideline that will fit all situations for grass carp. Each aquatic reservoir is different because of its own combination of fertility, water clarity, shallow water and chemical makeup. So, each of these variables affects the number of grass carp required to achieve the plant level to the desired control. Stocking rates may vary as low as one to as many as 20 grass carp per acre, depending on the amount and types of vegetation.

Stocking rates need to be increased as temperature decreases (as indicated by daily temperature units (DTU) decrease) because grass carp plant consumption and growth decrease. Stocking densities need to be based on the standing crop (biomass) of aquatic vegetation. This is estimated by multiplying plant distribution by average plant density; therefore, the higher the vegetation biomass, the higher the required stocking rate.

It should be well known that "overstocking" is followed by complete removal of all vegetation, while "understocking" of a water body causes either selective reduction of vegetation [21] or it can also result in no vegetation [22]. Low stocking densities can maintain intermediate plant control. On the other hand, plants rejected by the grass carp are left and may grow vigorously [23].

The amount of aquatic plants consumed by grass carp and its selectivity depends on many factors such as stocking density, age, temperature conditions, the length of time the fish have been in the pond and the quality and quantity of food present.

Initial plant density is an important indicator for the biological control. Biocontrol is effective if grass carp is stocked prior to the beginning of the rapid vegetation growth. Water level fluctuation should be estimated and taken into consideration. A dramatic decrease of water level could cause overstocking of the grass carp, and it is extremely difficult to remove fish from lakes. For this reason, stocking density of grass carp should be calculated for the lowest water level.

In addition to these, grass carp age and size are also important due to the possible predation on them, which can markedly reduce their initial stocking density. Grass carp should be larger than 30 cm when stocked; otherwise, they are very vulnerable to predators. In some areas, the otter can capture grass carp of about 2.7 kg (length of 60 cm), causing serious problems for fishpond management [24].

3.2. Changes in aquatic plant pattern and plankton composition

Grass carp can continuously control preferred aquatic plant species. Their impacts have been observed for 15–20 years at higher stocking rates. It is assumed that elimination of aquatic plant species preferred by the grass carp results in reduction of the diversity of the aquatic macrophyte community [25].

The stocking density and controlled plant area affect the extension of phytoplankton production in the ponds or lakes. In case of slow controlling of plants by grass carp, the indirect consequences of grass carp stocking on phytoplankton are negligible. It was determined that changes in the concentration of chlorophyll-a in the water were non-significant at low stocking density (30 kg ha^{-1}) [26]. Cassani et al. [27] also determined that in case of suppression of macrophytes, annual mean chlorophyll-a concentration remained stable in the ponds.

Primary production of the water reservoirs depends on light and nutrient availability. These two factors affect unstable equilibrium between macrophytes and phytoplankton. For this reason, the speed and extent of macrophyte removal by the grass carp affect the phytoplankton production.

Zooplankton consumption is necessary for juvenile and adult grass carp, but the consumed amounts are negligible in case the stocking density is not extremely high [28]. In lakes stocked with herbivorous fish, the growth of zooplankton and zoobenthos is enhanced through consumption of macrophytes by the fish and subsequently increased nutrient remineralization rates. The overall result can be also demonstrated through an increase of fish production [29]. Finally, the zooplankton communities shifted from copepod and copepod-cladoceran-dominated communities to rotifer and small cladocerans. Changes in zooplankton abundance and community structure were due to an increase in phytoplankton and shifts in planktivore predation on zooplankton by fish after macrophyte removal [30].

3.3. Changes in water quality and benthos

The effects of grass carp on plants and water quality are highly variable and often inconclusive due to the lack of proper control sites. The proportion and rate of plant removal by the grass carp is crucial. Changes in water quality as a result of plant removal by the grass carp mostly occur in small, non-flowing water bodies and least occur when only a small proportion of plants is removed from large, relatively deep, flowing reservoirs. In this concept, decreases can be observed in oxygen concentration of water following grass carp stocking, depending on the disappearance of macrophytes [31]. Primary producers such as phytoplankton and aquatic macrophytes not only release oxygen but also consume CO_2 during photosynthesis, which results in an increase in water pH. Changes in oxygen concentrations following grass carp stocking were positively correlated with the changes in pH [32].

Higher stocking densities of grass carp or their longer impact can increase concentrations of nutrients in the water, but these increases are mainly dependent on the water-body characteristics. These changes result from sediment resuspension during feeding and faecal matter deposition by carp as well as collapse of mechanisms responsible for maintenance of the vegetated state due to removal of macrophytes. Changes in benthos corresponded closely to changes in aquatic vegetation which stabilize sediments and provide additional substrate in the form of root masses and decaying material. Zoobenthos also responded to changes in water quality following removal of aquatic macrophytes [33].

The rate of aquatic plants elimination determines the magnitude of impact [30, 34]. These changes in water quality are often followed by algal blooms [35] which in most lakes signal a shift to an alternative stable state [36]. Increasing rates of nutrient cycling following resuspension of sediments lead to decreases in ecosystem stability [37].

4. Conclusion

In conclusion, grass carp can be effective in controlling of aquatic plants, but its potential adverse effects to aquatic ecosystems may be severe. In this concept, changes in plant abundance and community composition occur due to foraging activities, alteration of water transparency, disturbance of the sediment and deposition of faecal matter by grass carp. In addition, grass carp introductions may lead to unsuitable changes in the plant community. For this reason, risks and benefits of grass carp use should be considered, and necessary measures should be taken to control aquatic vegetation before stocking of grass carp to the aquatic environment.

Author details

Yusuf Bozkurt[1]*, İlker Yavas[2], Aziz Gül[3], Beytullah Ahmet Balcı[4] and Nurdan Coskun Çetin[2]

*Address all correspondence to: yfbozkurt@hotmail.com

1 Department of Aquaculture, Faculty of Marine Sciences and Technology, Iskenderun Technical University, Iskenderun, Hatay, Turkey

2 Department of Reproduction and Artificial Insemination, Faculty of Veterinary Medicine, Mustafa Kemal University, Antakya, Hatay, Turkey

3 Department of Animal Sciences, Faculty of Agriculture, Mustafa Kemal University, Antakya, Hatay, Turkey

4 Department of Aquaculture, Faculty of Fisheries, Akdeniz University, Antalya, Turkey

References

[1] Pitelli RA. Macrofitas aquaticas do Brasil, na condiçao de problema. In: Workshop Controle De Plantas Aquaticas. Resumos Brasília: Ibama; 1998. p. 19

[2] Zweerde VD. Biological control of aquatic weeds by means of phytophagous fish. In: Pieterse AH, Murphy KJ, editors. Aquatic Weeds: The Ecology and Management of Nuisance Aquatic Vegetation. Oxford: Oxford University Press; 1990. pp. 201-221

[3] Cross DG. Aquatic weed control using grass carp. Journal of Fish Biology. 1969;**1**:27-30

[4] FAO. The State of World Fisheries and Aquaculture [Internet]. 2014. Available from: http://www.fao.org/3/a-i3720e.pdf [Accessed: 15 March 2017]

[5] Shireman JV, Smith CR. Synopsis of Biological Date on the Grass Carp *Ctenopharyngodon idella* (Cuv. and Val., 1844). FAO Fish Synopses No. 135. Rome: FAO; 1983

[6] Chilton EW, Muoneke MI. Biology and management of grass carp (*Ctenopharyngodon idella,* Cyprinidae) for vegetation control: A North American perspective. Reviews in Fish Biology and Fisheries. 1992;**2**:283-320

[7] Page LM, Burr BM. A Field Guide to Freshwater Fishes. Boston: Houghton Mifflin Company; 1991. p. 432

[8] Eccles DH. FAO Species Identification Sheets for Fishery Purposes. Field Guide to the Freshwater Fishes of Tanzania. Rome: FAO; 1992. p. 145

[9] Opuszynski K, Shireman JV. Herbivorous fishes: Culture and use for weed management. In cooperation with James E. Weaver, Director of the United States Fish and Wildlife Service's National Fisheries Research Center. Boca Raton, Florida: CRC Press; 1995

[10] Keith P, Allardi J. Atlas des poissons d'eau douce de France. Patrimoines naturels. 1992;**47**:387. Paris: MNHN

[11] Beck RH. Risk assessment for the introduction of grass carp (*Ctenopharyngodon idella*). Report to the Canadian Non-Native Species Risk Analysis Committee. Northland Aquatic Sciences. Alberta, Canada; 1996. p. 94

[12] Goodchild CD. Non-indigenous freshwater fish utilized in the live food fish industry in Ontario: A summary of information. Ontario Ministry of Natural Resources. Ontario, Canada; 1999. p. 79

[13] Gorbach EI. Condition and fatness of the grass carp (*Ctenopharyngodon idella* Val.) in the Amur Basin. Journal of Applied Ichthyology. 1970;**11**:880-890

[14] Federenko AY, Fraser FJ. Review of grass carp biology. Interagency Committee on Transplants and Introductions of Fish and Aquatic Invertebrates in British Columbia. Fisheries and Marine Service Technical Report No. 786. British Columbia: Department of Fisheries and Environment; 1978. p. 15

[15] Stanely JG, Miley WW, Sutton DL. Reproductive requirements and likelihood for naturalization of escaped grass carp in the United States. Transactions of the American Fisheries Society. 1978;**107**:119-128

[16] NatureServe. NatureServe Explorer: An Online Encyclopedia of Life. Version 1.8. NatureServe, Arlington, Virginia [Internet]. 2004. Available from: http://www.natureserve.org/explorer [Accessed: 15 March 2017]

[17] Sutton DL, Vandiver VV, Hill J. Grass carp: A fish for biological management of hydrilla and other aquatic weeds in Florida. Florida Agricultural Experiment Station Bulletin. 2012;**867**:13

[18] Colle D. Grass carp for biocontrol of aquatic weeds. In: Gettys LA, Haller WT, Bellaud M, editors. Biology and Control of Aquatic Plants: A Best Management Practices Handbook. Marietta, Georgia: Aquatic Ecosystem Restoration Foundation; 2009. pp. 61-64

[19] Van Zon JCJ, Van der Zweerde W, Hoogers BJ. The grass carp, effects and side-effects. In: Proceedings of the 4th International Symposium on Biological Control of Weeds; Gainesville, FL. 1976

[20] Lopinot A. White amur, *Ctenopharyngogon idella*. Fish Management Mimeo No. 37. Illinois: Department of Conservation, Division of Fisheries; 1972. p. 2

[21] Blackwell BG, Murphy BR. Low-density triploid grass carp stockings for submersed vegetation control in small impoundments. Journal of Freshwater Ecology. 1996;**11**:475-484

[22] Bonar SA, Bolding B, Divens M. Effects of triploid grass carp on aquatic plants, water quality, and public satisfaction in Washington State. North American Journal of Fisheries Management. 2002;**22**:96-105

[23] Van Zon JCJ. Grass carp (*Ctenopharyngodon idella* Val.) in Europe. Aquatic Botany. 1977;**3**:143-155

[24] Adamek Z, Kortan D, Lepic P, Andreji J. Impacts of otter (*Lutra lutra* L.) predation on fishponds: A study of fish remains at ponds in the Czech Republic. Aquaculture International. 2003;**11**:389-396

[25] Catarino LF, Ferreira MT, Moreira IS. Preferences of grass carp for macrophytes in Iberian drainage channels. Journal of Aquatic Plant Management. 1997;**36**:79-83

[26] Pipalova I. Initial impact of low stocking density of grass carp on aquatic macrophytes. Aquatic Botany. 2002;**73**:9-18

[27] Cassani JR, Lasso-de-la-Vega E, Allaire H. An assessment of triploid grass carp stocking rates in small warm water impoundments. North American Journal of Fisheries Management. 1995;**15**(2):400-407

[28] Terrell JW, Terrell TT. Macrophyte control and food habits of the grass carp in Georgia ponds. Verhandlungen des Internationalen Verein Limnologie. 1975;**19**:2515-2520

[29] Zhang H, Chang WYB. Management of inland fisheries in shallow eutrophic, mesotrophic and oligotrophic lakes in China. In: Dudgeon D, Lam PKS, editors. Inland Waters of Tropical Asia and Australia: Conservation and Management. Stutgart, Vol. 24. 1994. pp. 225-229

[30] Richard DI, Small JW, Osborne JA. Response of zooplankton to the reduction and elimination of submerged vegetation by grass carp and herbicide in 4 Florida lakes. Hydrobiologia. 1985;**123**:97-108

[31] Opuszynski K. Impact of herbivorous fish culture on water quality in lakes. Pan'stwowa Inspecja Ochrony S'rodowiska w Zielonej Górze; 1997. p. 156

[32] Leslie AJ, Nall LE, Van Dyke JM. Effects of vegetation control by grass carp on selected water-quality variables in 4 Florida lakes. Transactions of the American Fisheries Society. 1983;**112**:777-787

[33] Gasaway RD. Benthic macroinvertebrate response to grass carp introduction in three Florida lakes. Proceedings of the Annual Conference on Southeastern Association of Fish and Wildlife Agencies. 1979;**33**:549-562

[34] Leslie AJ, Van Dyke JM, Hestand RS, Thomson BZ. Management of aquatic plants in multi/use lakes with grass carp (*Ctenopharyngodon idella*). Lake and Reservoir Management. 1987;**3**:266-276

[35] Klussman WG, Noble RL, Martyn RD, Clark WJ, Betsill RK, Bettoli PW, Cichra MF, Campbell JM. Control of aquatic macrophytes by grass carp in Lake Conroe, Texas, and the effect on the reservoir ecosystem. College Station, TX: Texas Agriculture Experiment Station MP-1664; 1988

[36] Scheffer M, Jeppesen E. Alternative stable states. In: Jeppesen E, Søndergaard M, Søndergaard M, Christoffersen K, editors. The Structuring Role of Submerged Macrophytes in Lakes. NY: Springer; 1997. pp. 397-406

[37] Wetzel RG. Limnology: Lake and River Ecosystems. 3rd ed. San Diego, CA: Academic Press; 2001. p. 1006

Section 3

Different Grass Species from Diverse World Regions

Chapter 4

Study of Some North African Grasses (*Ampelodesma mauritanica* and Esparto Grass)

Maghchiche Abdelhak

Additional information is available at the end of the chapter

http://dx.doi.org/10.5772/intechopen.70001

Abstract

Ampelodesma Mauritanica, commonly called Diss in Arabic very fibrous, is a plant of the family Poaceae, a plant native of northern Africa and southern Europe, is perennial and luxuriant, growing spontaneously in the wild state. Esparto grass in arid and semi-arid regions of North Africa; fight against the turning into desert regions form and was an important contributor in animal grazing and paper making. Natural fibers from plant are nowadays increasingly employed for replacing the synthetic materials due to economic and/or environmental considerations. A. Mauritanica and esparto grass fibers are cellulose-based fibers extracted using alkaline procedure to remove noncellulosic substances such as pectin, lignin, and hemicelluloses. The characterization of extracted fibers from both grass was based on the measurement of the morphological structure, chemical analysis, infrared spectroscopy, X-ray diffractometry, thermal analysis. The use of the natural fibers for composite materials, exhibits many benefits as it is low in weight, ecologically biodegradable, renewable, and cost-effective; it may have a role in local sustainable development in north Africa countries by valorizing these grass. Therefore, these grass materials could be a worthwhile choice for cellulosic fiber supply, and can lead to different useable products in order to improve the grass fiber added value.

Keywords: *Ampelodesma mauritanica*, esparto grass, fibers, chemical treatment, fiber extraction, cellulose

1. Introduction

Decreased and depleted petroleum resources coupled with augmentation of environmental problem through the world provides the alternatives for new biomaterials that are compatible with the environment and their development is independent of petroleum-based resources. A composite materials made from natural fiber reinforced with biodegradable polymer are in growing constantly the use of environmentally friendly materials [1].

Currently, numerous research groups have explored the production and properties of biocomposites where the polymer matrices are derived from renewable resources. They are the materials that have the capability to fully degrade and compatible with the environment.

At present, cellulose is the most abundant polymer available worldwide and a representative of renewable resources. In recent years, a number of bast-extracted fibers, alternative to the most used ones, such as jute, flax, hemp and kenaf, have been also proposed as reinforcement for plant fiber composites; these fibers are mainly from herbaceous plants. In general, trying to broaden the number of botanical species from which fibers are extracted may present interest.

The main purpose of this work is to characterize *Ampelodesma mauritanica* and esparto grass fibers in order to use them as reinforcement for structural composite materials. This choice is supported by the multiple advantages of natural fibers: they are available, renewable, and biodegradable, and they have a low price and represent an economic interest for the agriculture sector and different industries for different types of applications such as textiles, automobile constructions, medicines, paper industry and develop new products with vision and aim of sustainability.

1.1. Ampelodesma mauritanica (Diss)

A. mauritanica (Diss), family of Poaceae, is a plant native of northern Africa and southern Europe and is perennial and luxuriant growing spontaneously in wild state around the Mediterranean basin. The antiparasitic property is the only traditional use of this plant [2]. This plant previously was used in the realization of the old homes of North Africa regions because of its mechanical qualities. Possibly the use of such a fibrous plant in cements paste offers resistances very interesting, which make this material as an excellent filling lightweight for structures subjected to seismic [3]. Antibacterial and antifungal activities of alcohol extracts of aerial parts of Diss plant were examined. Phytochemical analysis shows that this plant rich in flavonoids and saponins, which might be responsible for its antimicrobial activity.

The hypoglycemic effect and antioxidant activity of the methanol extract of *A. mauritanica* roots were studied.

The levels of the total phenolic and the total flavonoid content in *A. mauritanica* roots, and also its antioxidant capacity were determined. The Methanol root extract could be a valuable source of hypoglycemic compounds; the phytochemical screening revealed the presence of flavonoids, saponins, cardenolides and tannins. It is observed that in some regions of Algeria, *A. mauritanica* has been used for reducing the blood glucose levels in diabetics [2, 4].

1.2. Esparto grass (*Stipa tenacissima* L.)

Stipa tenacissima L. (esparto grass or alfa grass) belongs to the Graminacies family; the scientific classification is shown in (**Table 1**); esparto grass is a perennial tussock grass widely distributed in semi-arid ecosystems of the southern and western Mediterranean basin. The esparto grass is a typically Mediterranean perennial herb, which grows in clumps about 1m to 1m20 high, thus forming large sheets. It grows spontaneously. Particularly in arid and semi-arid environments, it delimits the desert, where the esparto grass (alfa) stops, the desert begins [5]. It has a short fiber length which produces a paper of bulk; esparto wax has been a by-product of esparto used in furniture and boot polish manufacture.

Kingdom: Plantae
Order: Poales
Family: Poaceae
Genus: *Macrochlora* or *Stipa*
Species: *M. tenacissima*
Binomial name: Macrochloa tenacissima

Table 1. Scientific classification of esparto grass [6].

Esparto grass includes an underground part and an aerial part. The underground part, called the Rhizome, is formed as a complex network of highly branched roots about 2 mm in diameter and about 30–50 cm deep, ending in the young shoots (**Figure 1**).

The aerial part consists of several branches carrying cladding in the others, surmounted by limbs ranging from 30 to 120 cm in length.

The underside of the limbs is slightly shiny and the upper side carries strong ribs. Both are covered with an insulating wax which allows the plant to withstand drought.

The stem is hollow and cylindrical, and regularly interrupted at the node by entanglements of the bundles. The leaves are cylindrical, very tenacious, 50–60 cm long. The flower is protected by two glumes of equal length. The upper glume seems to be partially separated into two parts and the lower glume is finer. Generally, flowers appear in late April and early May and are green in color. The fruit is a caryopsis (a kind of grain) that is 5–6 mm long. Its upper part is brown and often carries dried traces. Fibers from leaves and stems are very strong and used in making paper; also the plant is a source of vegetable wax [7].

The Alfa, which is the Arab name of esparto grass (*Stipa tenacissima*) plant, is a hardy perennial grass from the family of the grass.

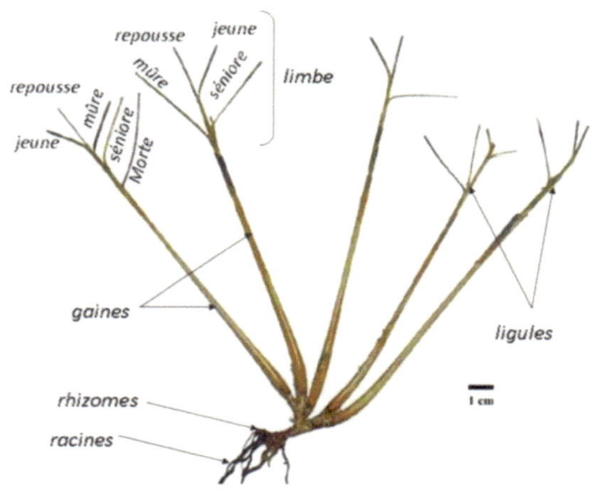

Figure 1. Morphology of esparto grass.

It is constituted of stems with a cylindrical shape which have a maximum height of about 1 m. Fiber differentiation is closely related to the vegetative cycle. The fibers are short and wide at the leaf base (basal level) and grow longer and thinner above the leaf [8]. The alfa stem is built of strong, stiff, and light cellulosic fibers which are mostly used in the production of high quality papers and for decoration, cigarettes, and dielectric applications for condensers; the stem is also used traditionally to manufacture ropes and carpets. The more recent information estimated that the esparto covered surface in hectares was approximately 3 million in Algeria [9, 10].

Due to its short fiber length, paper from esparto grass retains its bulk and takes block letters well. The esparto grass seems to prefer the calcareous soil, not very deep and permeable, with texture dominated by high sand rate. The high quantity of cellulose fiber contained in this plant, the flexibility, the smoothness, and the mechanical resistance of its fibers confer to it the very required properties in paper making; qualities recognized for a long time [11]. In addition to their multiple uses, natural fibers have shown many efficient properties such as heaviness, resistance, or flexibility which give them wide range of applications in textile field. In fact, they are recyclable and nature-friendly and nowadays they are exploited in automobile and medical applications. The fiber wastes from esparto grass offer a certain potential of liquid absorption and may be used as an absorbent fiber in hygienic products even in blends with fluff pulp [12]. In recent years, considerable attention has been given to the development and utilization of natural fibers. The main application of these materials as composites has been directed toward the automotive industries [13]. Composite materials were prepared using unsaturated polyester resins reinforced by Alfa fibers. Esparto grass fibers were evaluated for the production of bleached pulp [14]. To propose their application in composite materials, the question of chemical treatment is crucial, a preliminary study of chemical treatments to improve the properties of composite laminates [15]. As a consequence, a profound need exists for a sounder investigation of the morphological modifications produced by a wide range of chemical treatments on esparto fibers. Chemical treatments on esparto fibers increase the number of reaction sites [16], by removing non-structural matter such as hemicelluloses, lignin, and pectin [17]. However, alkaline treatment has been recognized capable of regenerating cellulose by the addition of OH groups, dissolving microscopic pits or cracks on the fibers alkali-treatment with NaOH and bleaching with sodium hypochlorite (NaClO) are applied on the fiber bundles to obtain the technical fiber used in textile products [18]. In North Africa, the esparto grass constitutes an essential element of fight against the turning into the desert and is an essential factor in the maintenance of balance in pastoral areas, due to well-developed root system that retains and protects the ground [19].

2. Materials and methods

2.1. *A. mauritanica* plant material

The Diss (*A. mauritanica*) is a very luxuriant plant. It is a large grass widespread in north Mediterranean Africa and the dry regions of south of Greece and Spain growing spontaneously in a wild state (**Figure 2**).

It is a high perennial herb. It blooms between April and June; the leaves are resistant. Growing up to a length of 1 m and 7 mm width, the plant is extremely rough and is used as a braiding

Figure 2. *Ampelodesma mauritanica* plant (Diss).

and serves as a material for paper [20]. Furthermore, recent studies have shown that it can be used as an additive to reinforce concrete [21]. It is also used by cattle breeders as antiparasitic and traditional medicine as antidiabetic [22]. Raw *A. mauritanica* stems (**Figure 3**) were collected from Batna in the east of Algeria during September 2016. The treatment started by purifying all of the Diss stems after removing the yellow and violet limbs. The aerial parts were air dried in shade at room temperature, then washed with water and Javelle water (12°) and then crushed [23]. The proportions of fiber contents were determined by treating stems with soda and sulfuric acids concentrate to obtain lignin, the cellulose obtained by bleaching and KOH treatment, the filtrate obtained is adjusted with acetic acid to obtain hemicelluloses.

Figure 3. Raw *Ampelodesma mauritanica* stems.

2.1.1. *Taxonomy of A. mauritanica*

- Reign: Vegetal
- Branch: Magnoliophyta
- Class: Liliopsida

- Order: Cyperales
- Family: Poaceae
- Genre: Ampelodesma
- Botanical name: *Ampelodesma mauritanica*

2.1.2. Pretreatment of A. mauritanica

The stems of *A. mauritanica* (**Figure 3**) was first rinsed with distilled water to remove dirt on the fiber surface. Washed *A. mauritanica* stems were left to dry at room temperature and finally dried in the oven for 5 h at 60°C. Afterward, they are submerged in 35 g/l salt water for 24 h at 60°C or 12 h to dissolve the waxes, a layer on the surface protects the plant against heat by limiting the evaporation of water.

2.1.3. Chemical characterizations of A. mauritanica

2.1.3.1. Water and volatile content

Water and volatile matter correspond to the loss of mass undertaken by the sample after drying in an oven at 100°C until constant weight (for 4 hours), water content and volatile matter (denoted by w) is expressed as:

$$W = \frac{m_1 - m_2}{m_1 - m_0} \times 100 \tag{1}$$

where m_0 (g): mass of empty crucible; m_1 (g): mass of crucible and test portion before heating; and m_2 (g): mass (crucible + residue) after heating up to constant weights.

2.1.3.2. Dry matter

The dry matter is determined from a raw sample, which is introduced into ceramic crucible, then weighed (m_0) and placed in the oven at 105°C until a constant weight. After cooling in a desiccator, the crucible containing the material dry is weighed (m_1).

The moisture content is then obtained from the equation below:

$$M\% = [(m_0 - m_1)/m_0] \times 100 \tag{2}$$

where m_0 = 2 g of Diss sample was placed in the oven at a temperature of 103–105°C for 4 hours. Then, the sample was allowed to cool in a desiccator containing $CaCl_2$ and the sample obtained was weighed.

2.1.3.3. Mineral content

Content of mineral substances (including mineral ash) is determined, by calcination of the dry sample at 500°C until constant weight. The calcined residue obtained is weighed.

The content of mineral matter (M_m) is expressed by:

$$M_m = \frac{m_3 - m_0}{m_1 - m_0} \times 100 \tag{3}$$

where m_3 is the weight of the crucible and of the residue after calcination to constant weight (g) and the difference of weight between the mass of dry matter and the mass of mineral content corresponds to the mass of organic matter.

2.1.4. Chemical extraction of A. mauritanica with NaOH

Ten grams of Diss sample was put in sufficient quantity of water for 2 hours and then the samples were dried after filtration. The sample was put in 1 l of NaOH (2 wt.%) and then put in water bath at 80°C for 2 hours, on filtrate and then washed with distilled water until the neutralization of pH, the processes were repeated twice.

2.1.4.1. Bleaching

The samples obtained from NaOH extraction (brown color) were bleached using sodium hypochlorite and acid buffer solution (27 g of soda in 50 ml of distilled water and 75 ml of acetic acid completed to 1 l with distilled water). The sample was put in a solution of H_2O_2/acetic tampon/NaClO (3:1:1). It is processed in a water bath at 80°C for 2 hours, then filtered, and washed twice until a white pulp of holocellulose was obtained.

2.1.4.2. Extraction of cellulose

The holocellulose obtained was dissolved in 70 ml of KOH (24 wt.%); the mixture was kept under stirring for 15 hours; then the paste was filtrated and washed until elimination of all KOH with distilled water; second wash was done with diluted acetic acid and then with ethanol; and finally, the cellulose was obtained dried and then weighed (**Figure 4**).

2.1.5. A. mauritanica spectroscopic analysis

Analysis of extracted cellulose fibers by infrared spectroscopy (FTIR-ATR) FTIR 8300 Fourier transformed SHIMADZY, allowed us for qualitative analyses to determine the groups functional properties present in the fibers. The results obtained are presented in **Figure 5**.

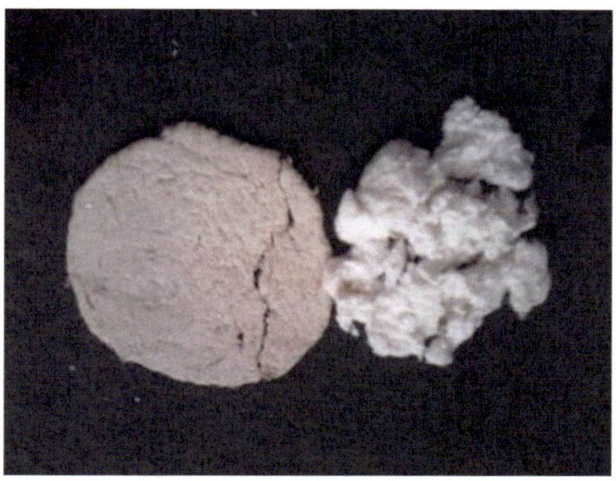

Figure 4. *Ampelodesma mauritanica* after chemical treatment.

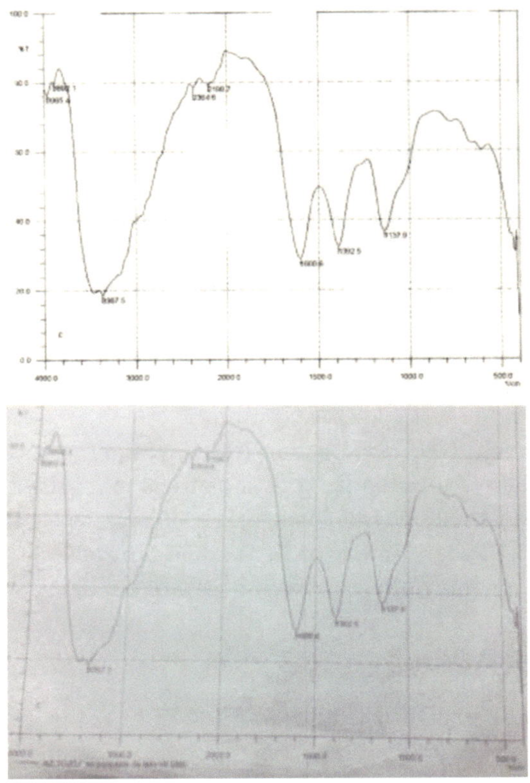

Figure 5. FTIR spectrum of *Ampelodesma mauritanica* fibers.

2.1.6. *A. mauritanica* X-ray diffraction

The crystalline structure (crystallinity rate) as a function of the chemical treatment studied by X-ray diffraction shown in **Figure 6,** was carried out on samples mechanically crushed using

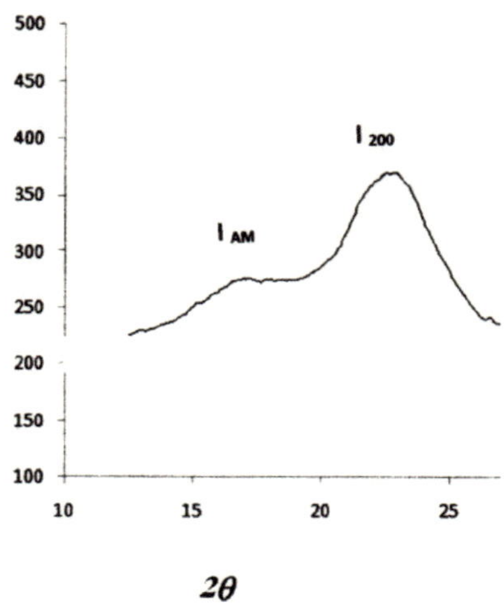

Figure 6. X-ray diffractogram of raw *Ampelodesma mauritanica* (Diss).

an electric mill, sieved to a size less than 125 μm. The diffraction was carried out by a Bruker D8 apparatus. The scanning speed is 0.2 s/step. The angle of diffraction was taken between 5 and 75°.

The monochromatic incident beam is centered on the copper Kα1 line (λ = 1.5418 A). The crystalline structure of plant fiber was studied using X-ray; the alkaline treatment of fibers improved the percentage of crystallization of cellulose in fiber. In order to ensure this, the crystallization factor was calculated according to the following equation [24]:

$$C = 100 \cdot \frac{I_{200} - I_{\text{non-cr}}}{I_{200}} \quad (4)$$

2.2. Esparto grass plant material

Esparto grass shown in **Figure 7**, is a hardy perennial grass of the grasses family, grows abundantly in dry, sunny locations along the seacoast. It bears gray-green leaves up to 1 m long and when young, provides food for cattle. Once esparto grass is mature, it becomes very tough. Esparto grass is an abundant plant in Algeria and it is used as paper paste with a low added value. The high percentage of cellulose fiber (41, 5 %) in this vegetal, the flexibility, the smoothness and the mechanical resistance of its fibers confer to him properties very required in paper making; qualities recognized since long time the esparto constitute a natural barrier which limits the expansion of the desert. The esparto grass seems to prefer the calcareous soil dominated by high sand rate. Leaf fibers are fibers that run lengthwise through the leaves of most monocotyledonous plants, such as esparto, these fibers, which are also referred to as hard fibers, are most commonly employed as reinforcing agents in plastics.

2.2.1. Pretreatment of esparto grass

The raw material shown in **Figure 8**, having been the subject of our study, comes from the Algerian steppe region.

Plant materials must be clean and free of extraneous substances, including soil and dust particles that may influence analytical results. For analyses of esparto grass, 10 g of finely crushed

Figure 7. Esparto grass (esparto, *Stipa tenacissima* L).

Figure 8. Raw esparto grass (esparto, *Stipa tenacissima* L).

plant was prepared approximately with particles of homogeneous size, sifted on sieve n° 24 and n° 27.

2.2.2. Organization and growth of the esparto grass

Esparto grass prefers the clay and silicic soil with a high rate of calcareous particles (30–40%) and a small percentage of gypsum (~2%) for its development [25]. Esparto tuft grows in a circle. The roots are at the stem's bases generally, they are not very deep. At the armpits of internodes appear the sheets, the buds, and the outlines of the future secondary roots (**Figure 9**), the stem is presented in the form of a thin ribbon, smooth, shining, solid, covered at the base with a hairy sheath.

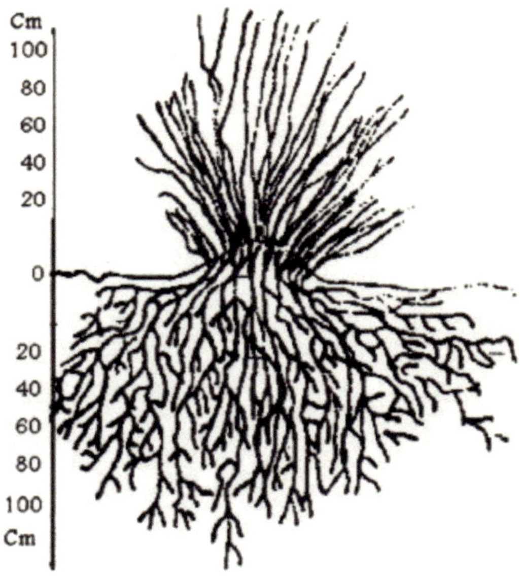

Figure 9. Radicular system of the esparto tuft.

2.2.3. Importance of esparto grass in desert regions

Esparto grass steppe habitats are important because of their ecological protection against desertification and economic (paper pulp manufacturing) interest, both on a national and international scale.

2.2.4. Esparto grass extraction procedure

First of all, the stems are being carded mechanically to refine their diameter. Before specific treatments were held, the fiber was first rinsed with distilled water to remove dirt from the fiber surface. Washed esparto fibers were left to dry at room temperature and finally dried in oven for 5 h at 60°C. Afterward, they are submerged in 35 g/l salt water for 24 h at 60°C or 12h at 80°C, to dissolve the waxes, a layer at the surface to protect the plant against heat by limiting the evaporation of water.

Esparto grass stems are treated with chemical products to degrade and eliminate the two main linking components, lignin and pectin. As the objective is to produce fibers, the hemicellulose does not need to be eliminated because it sticks the cellulosic filaments together to form fibers. Alfa fibers are cellulose-based fibers extracted from the esparto grass.

The cellulose was extracted from the esparto grass plant with 400 ml toluene/ethanol mixture (2/1, V/V) for 6 hours using a Soxhlet apparatus and treated with NaOH (1M) for 8 hours at 25°C [26], after filtration the cellulose was obtained and the filtrate contains the lignin and hemicellulose. This is mainly due to the reduction of lignin that binds the cellulose fibrils together.

2.2.5. Esparto grass plant analysis

The concentration of nutrients in Esparto plant tissues was measured in a plant extract obtained from fresh plant material, the esparto grass samples were washed in distilled water; oven dried at 60°C for 48 h, weighed, and then ground to 0.1 mm before chemical analysis. Elemental analysis was used to determine the organic, mineral and dry matter in the plant.

2.2.6. Esparto grass elemental analysis

Fibers extracted from esparto grass was characterized by different analysis methods.

2.2.7. Esparto grass characterization

Elemental analysis was used for analysis of esparto grass for its organic and mineral composition, the results for material composition of esparto grass are summarized in **Table 4** and the mineral components of esparto grass ashes are shown in **Table 5**.

2.2.8. Esparto grass FTIR analysis

A Perkin-Elmer 500 FT-IR spectrometer was used for infrared spectroscopy analysis of fibers extracted by KBr pellets.

2.2.9. *Esparto grass thermal analysis*

Thermal analysis (thermal gravimetric analysis [TGA] and differential scanning calorimetry [DSC]) was done for esparto grass fiber samples using Mettler TA TC 11 thermal analyzer.

2.2.10. *Esparto grass X-ray diffraction (XRD)*

A PW 1830 diffract meter was used to obtain the diffractogram of esparto grass fiber and the crystallinity of fibers was determined.

2.2.11. *Esparto grass morphological characteristics*

The dimensions of the fibers, and especially the length, are largely related to the quality of the Pulp [27], a manual method was used using a microscope equipped with an ocular micrometer. The average fiber size characteristics of the sample analyzed are summarized in **Table 6**.

2.2.12. *Esparto grass scanning electron microscopy (SEM)*

SEM micrographs were taken using Philips XL20 (Philips analytical Inc., the Netherlands). Samples were coated with gold before the examination (cathode dispersion). Fibers obtained from the esparto grass plant composed mainly of cellulose filaments were characterized by SEM.

3. Results and Discussion

3.1. *A. mauritanica* plant

3.1.1. *A. mauritanica plant analysis*

After extraction and bleaching of *A. mauritanica* plant, the cellulose fibers were obtained, fibers are better separated from one another, this causes the release of fibers encrusting substances (lignin, hemicellulose, pectin, etc.) and the chemical composition of *A. mauritanica* is shown in **Table 2**. Cellulose is the major component, followed by hemicelluloses and lignin, the smallest components are extractives and ashes.

3.1.2. *Mineralogical analysis of ashes after calcination of the A. mauritanica stem*

The mineral analysis (**Table 3**) of dissolved fiber ash was done using Atomic Absorption A.A-6200 SHIMADZU5.

***Ampelodesma mauritanica* (Diss)**	**Wt.% for each component**
Cellulose	41.5
Hemicellulose	22
lignin	27.5
Fats and waxes	1.3

Table 2. Percentage of each component of extracted *Ampelodesma mauritanica* stem.

Elements	SiO_2	CaO	MgO	K_2O	Na_2O	Fe_2O_3
%	72.25	4.50	1.45	3.60	0.70	2.31

Table 3. Ashes mineral components of *Ampelodesma mauritanica*.

Chemical composition of *A. mauritanica* plant ash results in its very variable mineral composition. The silica is very much present in the composition of the plant; this is even one of the reasons for which the delignification of this grass is carried out by the alkaline processes.

3.1.3. A. mauritanica FT-IR spectra

In FT-IR spectrum of *A. mauritanica* plant fiber (**Figure 5**), a broad absorption band at 3367.5/cm is mainly due to OH groups in the existing structure of the fibers. We also note the presence of a band at 1137/cm, and a second one at 1600/cm which indicates the existence of single C–O and double C═O bonds. At 1392 cm^{-1} liaison C-H and CH_2 deformation vibration, pic at 2340/cm Indicate the presence of OH acid carboxylic [28].

3.1.4. A. mauritanica X-Ray analysis

Figure 6 shows a diffractogram of raw *A. mauritanica* (Diss) fibers, in which one can observe that *A. mauritanica* fibers extracted via the physical-chemical process have a very similar diffraction pattern. Crystalline peak appears at I (200) corresponding to the intensity diffracted at $2\theta = 22.7°$. It is understandable that the cellulose content increases, whereas the amorphous hemicellulose content decreases during the physical-chemical process. I (AM) corresponds to the intensity diffracted at $2\theta = 15.2°$, crystallinity index of the crude *A. mauritanica* fibers is of the order of 34.27%, it is higher than that of the plant fibers of esparto grass which is 25% [29].

3.2. Esparto grass plant

3.2.1. Esparto grass plant analysis

After the extraction and bleaching process of esparto grass, cellulose fiber was obtained (**Figure 10, Tables 4** and **5**).

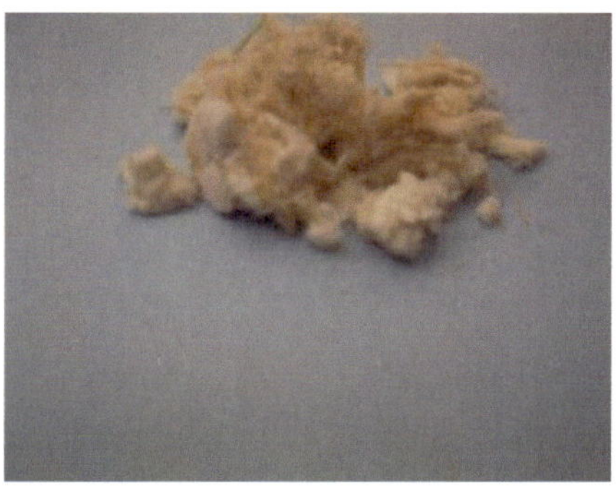

Figure 10. Cellulose fiber obtained from esparto grass.

Composition	Percentage of dry plant
Dry matter	94.25
Organic matter	17.78
Mineral matter	1.22
Extracted with ebullient water	4.06
Crude fiber	28.75
Cellulose rate	33.81
Lignin rate	18.2
Ash cotenant	5.75
Silica	2.03
Moisture	12.3

Table 4. Raw material composition of esparto grass.

Element	%
SiO_2	32.5
CaO	7.25
MgO	2.40
K_2O	1.32
Na_2O	0.40
P_2O_5	0.60
Fe_2O_3	2.60

Percentages of humidity of Esparto fibers absorption were found to be 67% at 25°C and losses on the ignition was 48.23% at 1100°C.

Table 5. Mineral components of esparto grass ashes.

3.2.2. Esparto grass IR spectra

In esparto fiber FTIR spectrum (**Figure 11**), a broad absorption band at (3274–3500)/cm is mainly due to groups OH in the existing structure of the fibers. We also note the presence of a band at 1050/cm, and a second to 1630/cm which indicates the existence of links C-O single and double C=O. Wave number 920/cm corresponds to the vibrations of -H aliphatic chains.

3.2.3. Esparto grass thermal analysis

Thermal stabilities and degradation patterns were determined by employing thermal gravimetric analysis (TGA) and differential scanning calorimetry (DSC). To examine the thermal

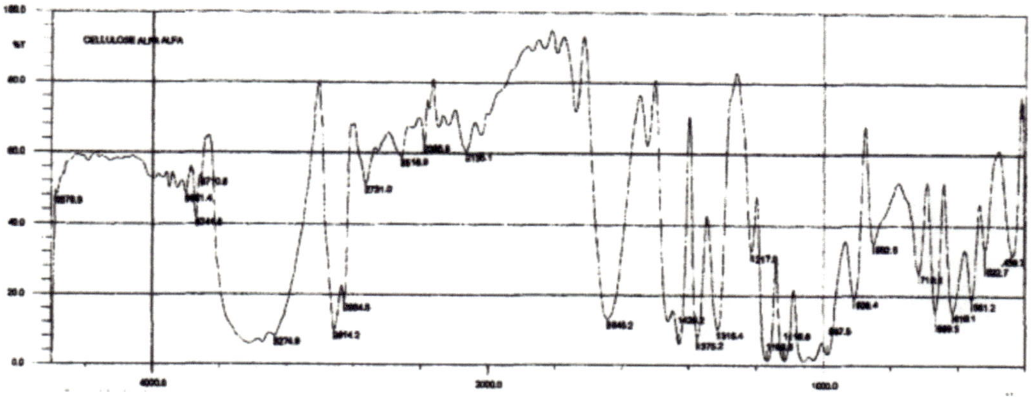

Figure 11. FTIR spectrum of esparto grass fibers with KBr.

stability of esparto grass fibers, thermogravimetric analysis under nitrogen flow was obtained.

A. Esparto grass fiber TGA curve

From esparto TGA curve (**Figure 12**), the weight loss at 70°C is due to the presence of H_2O in the sample and the degradation observed at 350°C.

B. Esparto grass fiber DSC Curve

Thermal degradation was studied by differential scanning calorimetry under nitrogen flow (**Figure 13**) and the cellulose shows one big endothermic curve at 80°C; caused by evolution of water entrapped by OH groups present in the cellulosic chains. The measured value indicate that the *decomposition of cellulose* begin at the *temperature* up than 200 °C.

3.2.4. Esparto grass X-ray analysis

From the X-ray diffractogram of extracted esparto grass fiber (**Figure 14**) the crystalline peak appears at 22.47°.

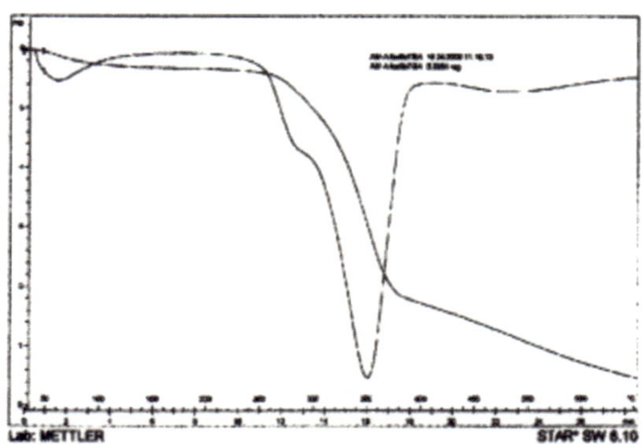

Figure 12. TGA curve of esparto grass fiber.

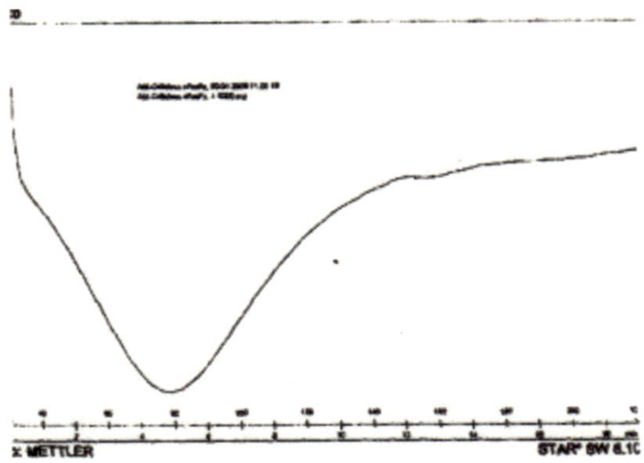

Figure 13. DSC curves of Cellulose Alfa-Alfa.

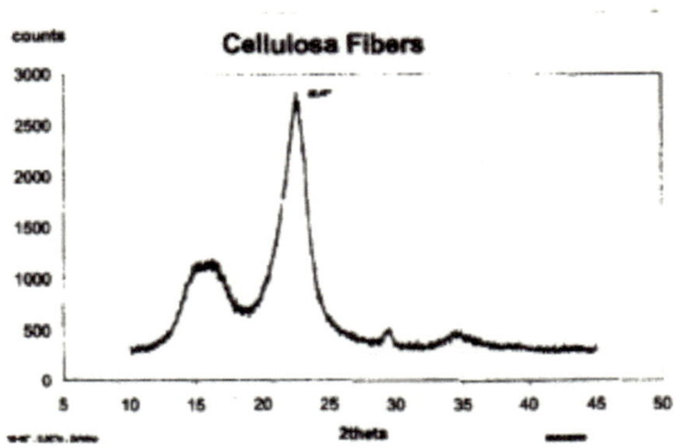

Figure 14. X-ray Diffractograms of esparto grass fiber.

This is understandable as the cellulose content increased, whereas the amorphous hemicellulose content decreased with the physico-chemical process [30]; this is in agreement with the FTIR and chemical analyses.

3.2.5. Esparto grass morphological characteristics

The average fibers sample dimensions analyzed are summarized in **Table 6**.

3.2.6. Esparto grass scanning electron microscopy (SEM)

The morphology of the esparto grass fibers was investigated by SEM as shown in **Figure 15**. Scanning electron micrographs of Esparto grass fibers and SEM micrographs (**Figure 16**)

Dimensions of esparto fiber	**mm**
Length	1.47
Outside diameter, *D*	0.015

Table 6. Dimensions of esparto grass fiber.

Figure 15. SEM of esparto grass fibers.

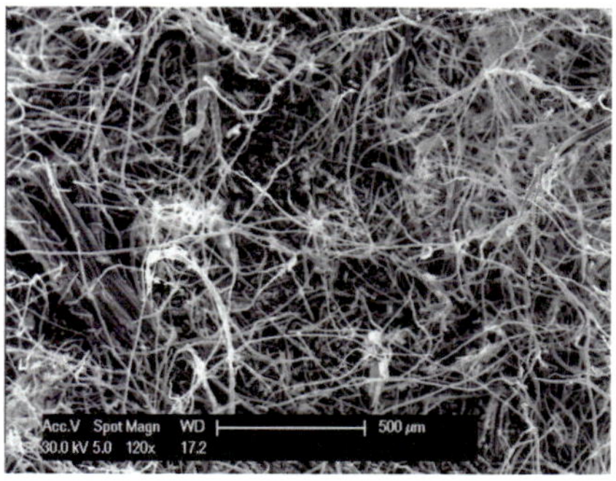

Figure 16. SEM of extracted esparto grass fibers.

of esparto grass fiber extracted, demonstrate that fibers are totally separated by chemical treatments.

4. Conclusion

This study aims to exploit new natural reinforcements for designing green composites. Cellulose fibers were extracted from *A. mauritanica* (Diss) and esparto grass (Halfa) plants for being used in different applications.

These materials are renewable, biodegradable, and very ecological. In fact, it requires a very small amount of water to grow and neither insecticides nor pesticides are needed.

Technical fibers have higher mechanical properties. However, ultimate fibers have a higher cellulosic rate.

To produce ultimate fibers, we must look for an appropriate method of extraction.

In this chapter, a method of extraction that gives cellulosic fibers without any damage was investigated; structure of *A. mauritanica* fibers is discontinuous, where cellulosic fibers are found in the matrix; this contains pectin, lignin, and hemicellulose. Fibers from esparto grass have a very short length and the structure of esparto grass fibers is discontinuous, where cellulosic fibers are found in the matrix, the obtained fibers are so fine and short. However, they have very interesting features. These fibers are then characterized through several analyses. The process of *A. mauritanica* and esparto grass fibers is simple and results in an excellent quality of fibers.

This study emphasizes the use of both fibers to different usable products in order to improve its added value.

The use of the natural fiber to prepare composite materials has major advantages like low price, low weight, biodegradable ecological, and renewable.

Author details

Maghchiche Abdelhak

Address all correspondence to: amaghchiche@yahoo.fr

Département de Pharmacie, Faculté de Médecine, University Batna 2, Algeria

References

[1] Sahari J, Sapuan SM. Natural fibre reinforced biodegradable polymer composites. Reviews on Advanced Materials Science. 2011;**30**(2):166-174

[2] Toudert N, Djilani SE, Djilani A, Dicko A, Soulimani R. Antimicrobial activity of the butanolic and Methanolic extracts of *Ampelodesma mauritanica*. Advances in Natural and Applied Sciences. 2009;**3**(1):19-21

[3] MERZOUD, Mouloud et HABITA, Mohamed Fouzi. Elaboration de composite cimentaire à base de diss «Ampelodesma Mauritanica». *Afrique Science: Revue Internationale des Sciences et Technologie*, 2008;**4**(2):231-245

[4] Djilani, A., N. Toudert, and S. Djilani. "Evaluation of the hypoglycemic effect and antioxidant activity of methanol extract of Ampelodesma mauritanica roots." *Life Sciences and Medicine Research* 31 (2011): 1-6

[5] Giménez, Gines Guzman. "Aportaciones a la química del esparto español." *Anales de la Universidad de Murcia (Ciencias)* (1955) : 342-346

[6] http://en.wikipedia.org/wiki/Esparto

[7] www.naturalmedicinalherbs.net/herbs/s/stipa-tenacissima=esparto-grass.php

[8] Belkhir S, Koubaa A, Khadhri A, Ksontini M, Smiti S. Variations in the morphological characteristics Of Stipa tenacissima fiber: The case of Tunisia. Industrial Crops and Products. 2012;**37**(1):200-206

[9] Brahim SB, Cheikh RB. Influence of fibre orientation and volume fraction on the tensile properties of unidirectional alfa-polyester composite. Composites Science and Technology. 2007;**67**(1):140-147

[10] Ghebalou A. Colloque sur l'industrie papetière. CIPA, Algeria. 2001

[11] Akechiche O, Messouad B. ХИМИЯ РАСТИТЕЛЬНОГО СЫРЬЯ, 2007, №4.С. 25-30

[12] Imed Ben Ali Ben merzoug TATAM. 2011;**7**(2):1-7

[13] Silva JLG, Al-Qureshi HA. Mechanics of wetting systems of natural fibres with polymeric resin. Journal of Materials Processing Technology. 1999;**92**:124-128

[14] Bouiri B, Amrani M. Elemental chlorine-free bleaching halfa pulp. Journal of Industrial and Engineering Chemistry. 2010;**16**(4):587-592

[15] Md M, Mohd M, Ma G, Santulli C. Lady's finger fibres for possible use as a reinforcement in composite Materials. Journal of Biobased Materials and Bioenergy. 2009;**3**(3):286-290

[16] Herrera-Franco P, Valadez-Gonzalez A. A study of the mechanical properties of short natural-Fiber reinforced composites. Composites Part B: Engineering. 2005;**36**(8):597-608. DOI: org/10.1016/j.compositesb.2005.04.001

[17] John MJ, Anandjiwala RD. Recent developments in chemical modification and characterization of Natural fiber-reinforced composites. Polymer Composites. 2008;**29**(2):187-207

[18] Msahli S, Sakli F, Drean JY. Study of textile potential of fibres extracted from Tunisian *Agave americana* L. AUTEX Research Journal. 2006;**6**(1):9-13

[19] Benaouda Z, Mehdadi Z, Bouchaour I. Influence pédoclimatique sur l'évolution des formations forestières En zone semi-aride (cas de la forêt de Tenira, Ouest algérien). Science et changements planétaires/Sécheresse. 2005;**16**(2):115-120

[20] Toudert N. de doctorat. Etude phytochimique et évaluation de quelques activités biologiques de *Ampelodesma mauritanica* [thèse]. Annaba: Université Badji Mokhtar; 2011

[21] Merzoud M, Habita MF. Elaboration of lignocellulosic composite formulated with a local resource: Diss as infill in structures submitted to seismic actions'. Research Journal of Applied Sciences, Medwell Journals. 2007;**4**(2):410-415

[22] Djilani A, Toudert N, Djilani SE, Dicko A. Hypoglycemic activity and antioxidant potential of *Ampelodesma mauritanica* extract. IUP Journal of Biotechnology. 2010; **4**(3) p55-62

[23] Sbiai A. Matériaux composites à matrice époxyde chargée par des fibres de palmier dattier: effet de l'oxydation Au tempo sur les fibres [doctoral dissertation]. Lyon: INSA; 2011

[24] Terinte N, Ibbett R, Schuster KC. Overview on native cellulose and microcrystalline cellulose I structure studied by X-ray diffraction (WAXD): Comparison between measurement techniques. Lenzinger Berichte. 2011;**89**;118-131

[25] Pouget M. Les relations sol-végétation dans les steppes sud-algéroises. 1980

[26] Heinze T, Dicke R, Koschella A, Kull AH, Klohr EA, Koch W. Effective preparation of cellulose derivatives in a new simple cellulose solvent. Macromolecular Chemistry and Physics. 2000;**201**(6):627-631

[27] Janin G. Technique de microcuissons papetières: Perfectionnement et précision des rendements en pâte. In annales des sciences forestières. EDP Sciences. 1981;38(1):107-126

[28] Mizi Fan, Dasong Dai and Biao Huang (2012). Fourier Transform Infrared Spectroscopy for Natural Fibres, Fourier Transform - Materials Analysis, Dr Salih Salih (Ed.), ISBN: 978-953-51-0594-7, InTech, Available from: http://www.intechopen.com/books/fourier-transform-materials-analysis/fourier-transform-infraredspectroscopy-for-natural-fibres

[29] Maghchiche, A., Haouam, A., & Immirzi, B. (2013). Extraction and characterization of Algerian Alfa grass short fibers (Stipa Tenacissima). *Chemistry & Chemical Technology*, (7, No 3), 339-344

[30] Reddy KO, Guduri BR, Rajulu AV. Structural characterization and tensile properties of *Borassus* fruit fibers. Journal of Applied Polymer Science. 2009;**114**:603-611. DOI: 10.1002/app.30584

Chapter 5

Endemic Species of the Family Poaceae in Chile: Taxonomy, Distribution, and Conservation

Víctor L. Finot, Alicia Marticorena,
Roberto Rodríguez and Romina G. Muñoz

Additional information is available at the end of the chapter

http://dx.doi.org/10.5772/intechopen.68973

Abstract

Due to its geographic isolation, Chile is a biogeographic island which harbors a high percentage of endemism. More than 50% of the native vascular flora is endemic and more than 60% lives in Central Chile, included in the Chilean Biodiversity Hotspot. Endemic are species with a geographic distribution restricted to a single area and could be especially vulnerable. For these reasons, updated lists of endemic species are necessary. Based on a databases, the study of specimens from two Chilean herbaria and the available literature, we present an updated list of grasses endemic to Chile indicating for each taxon the scientific accepted name, common names, type, life cycle, flowering period, distribution, conservation status, bibliographic references and representative specimen. Seventy-one species (19.9% of the native grass species) were classified as endemic. Most species occur along the Chilean hotspot of biodiversity, mainly in the Mediterranean region of the hotspot. One species (*Podophorus bromoides*) is extinct, three species are critically endangered, two species are endangered, one species is vulnerable, and one species is near threatened. The conservation status of most species (89%) needs to be evaluated. Most of the threatened species are endemic to the Juan Fernández Archipelago.

Keywords: checklist, conservation status, endemism, gramineae, herbarium

1. Introduction

Endemics are species with a geographic distribution limited to a single area (local, regional, national, or continental), regardless of the size of the area. Species that live in a narrow geographical area are commonly called *rare* species. If a species lives in a single and narrow geographic area, it is a rare endemic species and could be especially vulnerable. Rare endemic species that live in insular territories are especially prone to extinction. In this chapter, we consider endemics as those grass species restricted to Chile, comprising both continental and insular territories [1]. In Chile, a high percentage of the fauna and flora is endemic; on average, nearly 25% of the species are endemic. Some groups harbor a high percentage of endemic taxa: amphibians (65%), reptiles (63%), fishes (55%), and vascular plants (nearly 50%) [2]. Endemic species are of great biological importance because they provide unique genetic diversity [3]. In this chapter, we provide an updated list of the species of the family Poaceae endemic to Chile, based on the study of herbarium specimens and all available taxonomic literature.

2. Materials and methods

A database of the species of the endemic grasses of continental and insular Chile was constructed based on the databases of two Chilean herbaria: CONC (Herbarium of the University of Concepcion) and SGO (Herbarium of the National Museum of Natural History, Santiago). Specimens deposited in these herbaria and those collected for this project were studied. Data were supplemented with specimens cited in previous publications [4–8]. For each species, we include taxonomic data (scientific accepted names, synonyms, common names, and types of accepted names), life cycle, flowering period, distribution, conservation status, bibliographic references, and representative specimens.

3. Results

3.1. Taxonomic distribution

Of the 356 native species identified for the Chilean flora, 19.9% (71 species) were classified as endemic. The endemic Chilean grasses belong to five subfamilies and 27 genera. Most of the species belong to Pooideae (55 spp., 77.5%) from 20 genera, Bambusoideae comprises six endemic species (8.5%) of the genus *Chusquea*, Danthonioideae includes five spp. (7.0%) in two genera (*Danthonia* and *Rytidosperma*), Chloridoideae includes three spp. (4.2%) from two genera (*Cynodon* and *Eragrostis*), and Panicoideae two spp. (2.8%) from two genera (*Imperata* and *Paspalum*). Two species are endemic to Easter Island (*Rytidosperma paschale* and *Paspalum forsterianum*), one species (*Eragrostis kuschelii*) is endemic to San Ambrosio Island (Desventuradas Islands), and four species are endemic to the Juan Fernández Archipelago (*Agrostis masafuerana, Chusquea fernandeziana, Megalachne berteroana, Megalachne masafuerana*). *Podophorus bromoides* was described for Juan Fernández but it seems to be currently extinct [9]. The most diverse genera are *Nassella* (11 spp.), *Poa* (eight spp.), *Melica* (seven spp.), and *Chusquea* (six spp.).

3.2. Geographic distribution

In continental Chile, the number of species per Administrative Region varies from three in the regions of Tarapacá (20°S), Atacama (27°S), and Magallanes (53°S) to 26 in Coquimbo (29°S) and

Biobío (36°S). It is interesting to note that most of the species occur along the Chilean hotspot of biodiversity, mainly in the Mediterranean region of the hotspot (23–38°S). Moreover, this interesting area is the most populated zone of the country with a high degree of urbanization, agriculture, industrial activities, and forestry [7]. The number of endemic species in the northernmost regions (Arica and Parinacota, Tarapacá, Atacama) is fairly low as well as in the southernmost regions (Aysen, Magallanes). The same occurs in insular Chile. The Juan Fernández Archipelago contains seven species endemic to Chile, Easter Island two species, and Desventuradas Island only one species (**Figure 1**). The species more widely distributed are *Phalaris amethystina* (nine regions including Juan Fernández Island), *Melica violacea* (eight regions), and *Danthonia malacantha* (eight regions including Juan Fernández Island). Local endemic species are *Agrostis insularis* (Chiloé Island, Los Lagos), *A. masafuerana* (Masafuera, Juan Fernández Island), *Alopecurus lechleri* (Los Ríos Region), *Bromus burkartii* (Araucanía Region), *Chusquea fernandeziana* (Juan Fernández Island), *Eragrostis kuschelii* (Desventuradas Islands), *E. pycnantha* (Atacama Region), *Imperata parodii* (Araucanía Region), *Megalachne berteroana* and *M. masafuerana* (Juan Fernández Island), *Melica poecilantha* (Coquimbo Region), *Nassella coquimbensis* (Coquimbo Region),

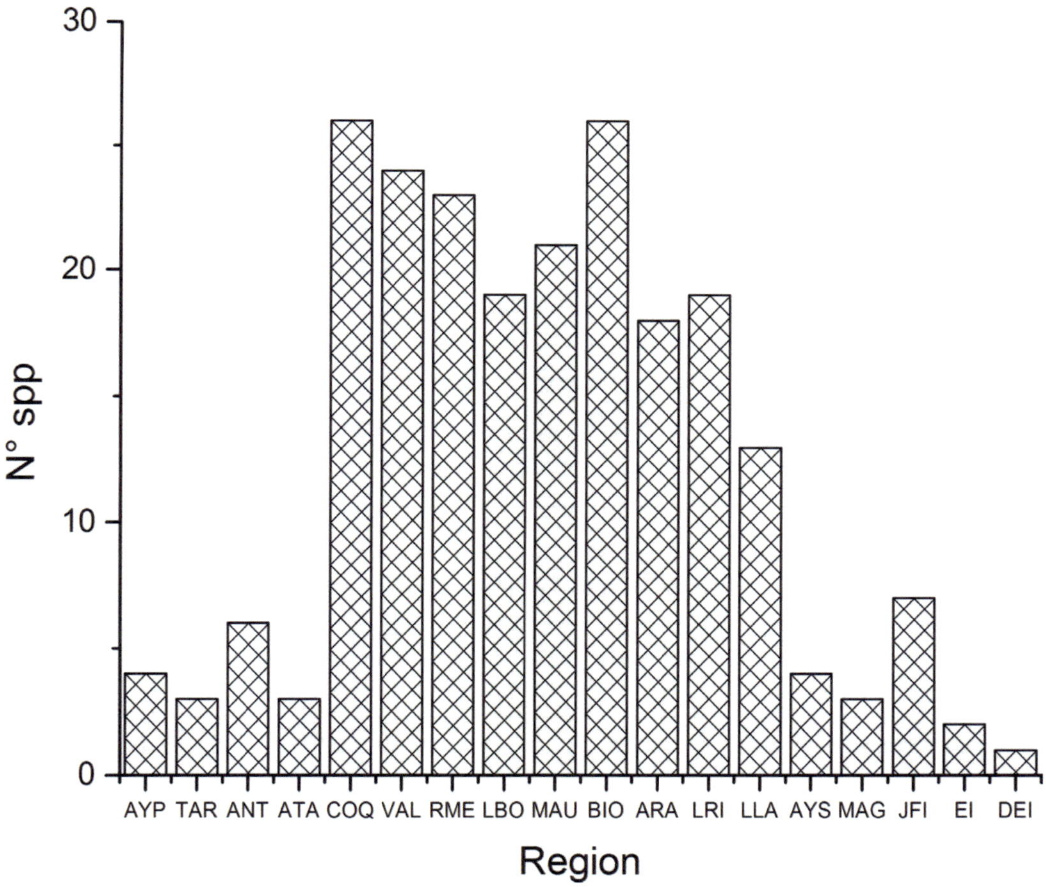

Figure 1. Number of endemic species in the 15 political regions of Chile.

Paspalum forsterianum (Easter Island), *Poa pfisteri* (Biobío Region), *Poa schoenoides* (Los Lagos Region), *R. paschale* (Easter Island), and the subspecies *Trisetum johnstonii* subsp. *mattheii* (Arica and Parinacota Region).

3.3. Conservation

One species (*P. bromoides*) is extinct, three species are classified as critically endangered (*A. masafuerana, C. fernandeziana,* and *M. masafuerana*), two as endangered (*E. pycnantha* and *Paspalum chilense*), one as vulnerable (*Megalachne berteroniana*), and one species as near threatened (*M. poecilantha*). The conservation status of most species (89%) needs to be evaluated. Moreover, several species are known only by their type collections. Most threatened species are endemic to the Juan Fernández Archipelago.

3.4. Updated list of the endemic Chilean grasses

1. ***Agrostis arvensis*** Phil., Linnaea 29(1): 87. 1858. **Typus**: CHILE "In arvis prope Concepción legit Cl. Gay et in herb. Chil. sub nomine *Poa* no 226 reliquit" (Holotype: SGO).

Life cycle: Short-lived perennial or probably annual [10]. **Flowering period**: January. **Distribution**: A rare endemic species distributed between 36°50′S, 73°03′W (Concepción) and 40°42′S, 71°57′W (Osorno), in the Regions of Biobío and Los Lagos, southern Chile, from 10 to 420 m; it was collected only near Concepción (Biobío), National Park Futangue (Los Ríos Region), and National Park Puyehue (Osorno, Los Lagos Region). **Conservation status**: Not evaluated. **References:** [10]. **Comments**: According to Muñoz-Schick [11], this taxon is a synonym of *Agrostis vidalii* Phil. growing in Chile and Argentina, and consequently, losing its endemic status. Other authors accept *A. arvensis* as a valid species, with its distribution restricted to the Biobío Region, whereas *A. vidalii* grows in the regions of Los Lagos and Aysen [12–14]. **Representative specimen**: Biobío Region, Concepción, Gay 217 (SGO).

2. ***Agrostis insularis*** Rúgolo & A.M. Molina, Gayana Bot. 54(2): 111, Fig. 10, 11. 1997. **Typus**: CHILE, X Región, Prov. Chiloé, Isla Alao, lado norte, barrancos, 30 m.s.m. (42°35′S–73°16′W), 15-I-1985, *C. Villagrán e I. Mesa 5890* (Holotype: CONC).

Life cycle: Perennial. **Flowering period**: January. **Distribution**: Restricted to the Chiloé Archipelago, between 42°35, 73°16′W, and 42°38′S, 73°16-W, from 25 to 30 m. It has been collected in Alao Island and Chaulinec Island. **Conservation status**: Not evaluated. **References:** [10, 12]. **Representative specimen**: Chile, Los Lagos Region, Chiloé, Isla Alao, 30 m, Villagrán & Mesa 5890 (CONC, type).

3. ***Agrostis masafuerana*** Pilg., Repert. Sp. Nov. Regni Veg. 16: 388. 1920. **Typus**: CHILE, "Juan Fernández, Masafuera, hochland bei Las Torres, ca. 1300 m, 14 Feb 1917", *C. & I. Skottsberg 424* (Holotype: B; isotypes: BAA, K, P, US).

Life cycle: Perennial. **Flowering period**: January to February. **Distribution**: Endemic of the Más Afuera Island (33°45′S, 80°47′W) (=Masafuera Island, Alejandro Selkirk Island), Juan Fernández Archipelago, at 800–1350 m.a.s.l. **Conservation status**: Critically endangered (CR).

References: [10, 12, 15–17]. **Representative specimen**: Valparaíso Region, Juan Fernández Archipelago, Masafuera Island, Cuchillo del Imán, Landero & Gaete 9155 (CONC).

4. ***Agrostis umbellata*** Colla, Herb. Pedem. 6: 18. 1836. **Typus**: CHILE "habitat Chili, Rancagua, 1828, *D. Bertero 31 & 556*".

Life cycle: Short-lived perennial or annual. **Flowering period**: October to January. **Distribution**: From Petorca, Region of Valparaíso to the Region of Aysen (Regions of Valparaíso, Metropolitan, O'Higgins, Biobío, and Aysen), from 900 to 2000 m. It grows in damp, sandy places. **Conservation status**: Not evaluated. **References:** [10, 12, 18]. **Comments**: This species was considered endemic to Chile by Rúgolo & Molina [10]. However, McCloskie cited it previously for Argentina in 1915 [19]: "*I have found it in a good many places in the western parts of the Sta. Cruz Territory, for instance at Lago Argentino, Lago Viedma and Lago San Martín…*" and by Spegazzini in 1921 [20]: "*In pratis montanis secus Carren-leofú, aest. 1899–900.*" However, as Rúgolo & Molina [10] stated, no reference specimens collected in Argentina can be found in herbaria. On the other hand, *Agrostis umbellata* is very close morphologically to *A. inconspicua* Kunze ex E. Desv., growing in Chile and Argentina. **Representative specimen**: Valparaíso Region, Quillota, cerro La Vizcacha, 33°05'S, 71°02'W, 2000 m, Zoellner 3325 (CONC).

5. ***Alopecurus heleochloides*** Hack., Repert. Spec. Nov. Reg. Veg. 10(243–247): 166. 1911. **Typus**: Chile, inter Colina et Batuco, Nov 1899, *C. Reiche s.n.*

Life cycle: Annual. **Flowering period**: September to December. **Distribution**. A rare endemic, known only from the Coquimbo Region, Choapa (31°35'S) and the Metropolitan Region, Santiago (33°12'S) from 25 to 3300 m.a.s.l. It has been also collected on saline soils in herbaceous steppe and in Andean Patagonia at the Region of Magallanes. **Conservation status**: No evaluated. **References:** [21–25]. **Representative specimen**: Coquimbo Region, Choapa, Mincha Sur, 5 Km E of Huentelaquen, 180 m, Bliss et al. 2139 (CONC).

6. ***Alopecurus lechleri*** Steud., Syn. Pl. Glumac. 1: 148. 1854. **Type**: CHILE, "hbr. *Lechler 440*, Valdivia."

Life cycle: Perennial. **Flowering period**: Unknown. **Distribution**: Known only from the Region of Los Ríos, Valdivia (39°50'S). **Conservation status**: Not evaluated. **References:** [22, 25]. **Representative specimen**: Region of Los Ríos, Valdivia, Lechler 440 (Type).

7. ***Anatherostipa venusta*** (Phil.) Peñailillo, Gayana Bot. 53(2): 279. 1996. **Basionym**: *Stipa venusta* Phil. **Typus**: Chile, Tarapacá, de Socaire allata, Feb 1885, F. Philippi. **Synonym**: *Nicoraella venusta* (Phil.) Torres. **Common name**: Vizcachera.

Life cycle: Perennial. **Flowering period**: September to May. **Distribution**: It grows in northern Chile, Regions of Arica and Parinacota, Tarapacá and Antofagasta, from 3400 to 5000 m.a.s.l. It forms dense cushions in arid sandy soils. **Conservation status**: Not evaluated. **References:** [26–29]. **Comments**: This species is closely related to *Anatherostipa boomani* (Hauman) Peñailillo from Chile and Argentina from which it differs by its glabrous floret (floret pubescent in *A. boomani*) [29]. **Representative specimen**: Arica and Parinacota Region, Tacora-Humapalca-río Azufre, 4300 m, Teillier 7759 (CONC).

8. ***Bromidium trisetoides*** (Steud.) Rúgolo, Darwiniana 24(1–4): 187–216. **Basionym**: *Agrostis trisetoides* Steud., Syn. Pl. Glumac. 1: 172. 1854. **Type**: "Hrbr. Lechler 724, Arigue [Arique, Valdivia], Chile."

Life cycle: Annual. **Flowering period**: October to January. **Distribution**: Regions of Biobío, Araucanía and Los Ríos, 75–300 (−1000) m.a.s.l. on sandy and humid soils [30]. Sometimes in temporary wetlands. **Conservation status**: Not evaluated. **Reference**: [30]. **Representative specimen**: Araucanía Region, 10 km S of Cholchol, 30 m, Bliss 2426 (CONC).

9. ***Bromus gunckelii*** Matthei, Gayana Bot. 43(1–4): 62. 1986. **Typus**: "Chile, I Región, Provincia Parinacota, frente a Socoroma, en quebrada surcada por riachuelo, 3300 m.s.m., Ricardi, Weldt & Quezada 192, 5-V-1972" (Holotype: CONC).

Life cycle: Annual. **Flowering period**: March to May. **Distribution**: Restricted to the Region of Arica and Parinacota and Region of Tarapacá, from 3100 to 3740 m.a.s.l. **Conservation status**: Not evaluated. **Reference**: [31]. **Comments**: *Bromus gunckelii* is closely related to *B. berteroanus* Colla. Both species are annuals with lax panicles and with a twisted awn borne near the apex of the lemma. These species differ mainly on the length of the glumes relative to the length of the adjacent lemmas. In *B. berteroanus,* the glumes are longer than half the length of the adjacent lemma, whereas in *B. gunckelii* both glumes are shorter than half of the adjacent lemmas. **Representative specimen**: Region of Arica and Parinacota, camino Arica a Portezuelo de Chapiquiña, Ricardi et al. 138 (CONC).

10. ***Bromus burkartii*** Muñoz, Agric. Técn. (Santiago) 8: 83. 1948. **Typus**: Chile: Prov. Cautín, Lonquimay, Cordillera de Las Raíces, habitante de las vertientes, 2 Mar 1939, *Burkart 9504* (HT: SGO).

Life cycle: Perennial. **Flowering period**: December to January. **Distribution**: This species grows only in the Region of the Araucanía where it has been collected in Lonquimay, Lolén, Galletué National Park, and Liucura under *Nothofagus pumilio* (lenga)-*Araucaria araucana* (pehuén) forest and on waysides. **Conservation status**: Not evaluated. **Reference**: [31]. **Comments**: *Bromus burkartii* was considered a good species by Matthei [31] but modern treatments [12, 13] consider it a synonym of *B. mango* E. Desv. from Argentina and Chile, losing its endemic status. *B. burkartii* differs from *B. mango* mainly by the size of the lemma (5.5–7.0 mm in *B. mango,* 8–11 mm in *B. burkartii*). **Representative specimen**: Araucanía Region, Lonquimay, Cordillera de Las Raíces, 1490 m, 7 Jan 1984, Matthei & Bustos 98 (CONC).

11. ***Chascolytrum koelerioides*** (Trin.) L. Essi, Longhi-Wagner & Souza-Chies, Novon 21(3): 328–329. 2011. **Basionym**: *Poa koelerioides* Trin., Mem. Acad. Imp. Sci. Saint-Pétersb., sér. 6, Sci. Math., Seconde Pt. Sci. Nat. 4: 62. 1836. **Typus**: V spp. Chil.

Life cycle: Perennial. **Flowering period**: October to January. **Distribution**: This species is common in central and central-southern Chile, from Valparaíso to Valdivia, from 5 to 1300 m of elevation (Regions of Valparaíso, O'Higgins, Maule, Biobío, Araucanía and Los Ríos). **Conservation status**: Not evaluated. **References**: [22, 32–34]. **Representative specimen**: Valparaíso Region, Quillota, Cerro La Campana, Puerta De Ocoa, Zoellner 13137 (CONC).

12. ***Chascolytrum rhomboideum*** (Link) Essi, Longhi-Wagner & Souza-Chies, Novon 21(3): 329. 2011. **Basionym**: *Rhombolytrum rhomboideum* Link., Hort. Berol. 2: 296. 1833 (as *Rhomboelytrum*). **Type**: Habitat in Chile e seminibus ideallitis enata, Anonymous.

Life cycle: Perennial. **Distribution**: Central Chile, Regions of Valparaíso, Metropolitan, O'Higgins and Maule. **Conservation status**: Not evaluated. **References**: [12, 33]. **Representative specimen**: Valparaíso Region, Quillota, Limache, Cerro Cruz, Garaventa 2523 (CONC).

13. ***Chusquea ciliata*** Phil., Linnaea 33(3–4): 299. 1864. **Typus**: Chile "Locis maritimis prov. Santiago, prope Algarrobo, etc., Julio florebat, incolis Quila".

Life cycle: Perennial. **Distribution**: Central Chile, Regions of Valparaíso and Metropolitan, associated to *Teline monspessulana, Alstroemeria* sp. and *Maytenus boaria*. **Conservation status**: Not evaluated. **References**: [12, 35–39]. **Comments**: A very uncommon species, related to *Ch. uliginosa*. McClure [39] considers *Ch. ciliata* a synonym of *Ch. tenuiflora* Phil. However, other authors [12, 37] treated it as a good species. **Representative specimen**: Region of Valparaíso, Viña del Mar, Rodelillo, 308 m, Novoa 267 (CONC).

14. ***Chusquea cumingii*** Nees, Linnaea 9: 487. 1834. **Typus**: Chile "Prope Valparaiso Regni Chilensis, 1831, H. Cuming et Bridges in Herbario Lindleyano." **Common names**: Colihue, coligüe, quila, quila chica, colihue de la zona central.

Life cycle: Perennial. **Distribution**: Central Chile, in the Regions of Coquimbo, Valparaíso, Metropolitan, O'Higgins, Maule and Biobío, from 15 to 1450 m.a.s.l., associated to *Trevoa trinervis, Cryptocarya alba, Centaurea chilensis, Nothofagus macrocarpa*, and *Beilschmiedia miersii*. **Conservation status**: Not evaluated. **References**: [22, 35, 37]. **Representative specimen**: Region of Valparaíso, Parque Ncional La Campana, mina Pronosticada, 1300 m, Looser s.n. (CONC).

15. ***Chusquea fernandeziana*** Phil., Anales Univ. Chile 43: 577. 1873. **Typus**: Chile, Hallé este coligüe en la Isla de Juan Fernández" Philippi s.n. **Common name**: Coligüe.

Life cycle: Perennial. **Distribution**: Endemic to the Juan Fernández Archipelago (Robinson Crusoe or Masatierra Island). **Conservation status**: Critically endangered (CR). **References**: [9, 16, 40]. **Representative specimen:** Region of Valparaíso, Juan Fernández Archipelago, Masatierra, Baeza & Peñailillo 11330 (CONC).

16. ***Chusquea macrostachya*** Phil., Anales Univ. Chile 94: 350. 1896. **Typus**: "Chile, pariter in valle fluminis Palena invenit orn. Fr. Delfin." **Icon.**: Muñoz Schick, 1980: 492. **Common name**: Taihuén.

Life cycle: Perennial. **Distribution**: Regions of Los Ríos and Los Lagos from 70 to 1100 m.a.s.l. where it is an important component of the *Nothofagus betuloides-Chusquea macrostachya* forest. **Conservation status**: Not evaluated. **References**: [22, 39, 41–44]. **Representative specimen**: Los Lagos Region, Chiloé, entre Chonchi y Quellón, Matthei 394 (CONC).

17. ***Chusquea montana*** Phil. fma. ***nigricans*** (Phil.) Matthei, Gayana, Bot. 54 (2): 204–205, f. 11. 1997. **Basionym**: *Chusquea nigricans* Phil., Anales Univ. Chile 27 (2): 323. 1865. **Typus**: Chile: Frequens in montibus litoreis prov. Valdiviae "Cordillera Pelada" dictis. *Philippi 518* (SGO). **Common name**: Quila enana.

Life cycle: Perennial. **Flowering period**: November. **Distribution**: Southern Chile, Regions of Los Ríos, Los Lagos and Aysen, from 300 to 1300 m.a.s.l. in peatlands or wetlands (mallines). **Conservation status**: Not evaluated. **References**: [12, 35, 39, 42, 45]. **Comments**: Philippi [46] characterized this species as having very short glumes; he notes that only *C. breviglumis* (=*C. culeou* E. Desv.) shares this character. He also highlights some characteristics of the leaf blade. Parodi [35] trated *C. nigricans* as synonym of *C. montana*. In his opinion, the types of both species differ by characteristics of the leaf blade and the inflorescence (*C. nigricans* has narrower and shorter leaf blades with more prominent veins and shorter inflorescences than *C. montana*). Because he found specimens with intermediate characteristics, he suggests that *C. nigricans* should be treated as synonym of *C. montana*. Later, Matthei [41] separates it from *C. montana* at the level of forma. This concept prevails in modern treatments [12]. **Representative specimen**: Region of Los Ríos, Valdivia, Cordillera Pelada, Ricardi & Matthei 5249 (CONC).

18. ***Chusquea uliginosa*** Phil., Linnaea 30 (2): 207. 1859. **Typus**: Chile "in pratis illis pluvias inundatis, uliginosis, quae incolis Nadi, provinciae Valdiviae; a colonis Germanis Kleine Quila vocatur".

Life cycle: Perennial. **Distribution**: This species grows in Southern Chile, in wet places from the Regions of Los Ríos, Los Lagos, and Aysen. It grows up to 1500 m.a.d.l. associated to *Nothofagus alpina, N. antarctica, N. obliqua,* and *Berberis valdiviana,* sometimes in peatlands. **Conservation status**: Not evaluated. **References**: [22, 35, 39, 42]. **Representative specimen**: Los Lagos Region, Osorno, Antillanca, Schlegel 7016 (CONC).

19. ***Cynodon nitidus*** Caro & E.A. Sánchez, Darwiniana 17: 510, f. 1. 1972. **Typus**: "Chile, Dpto. Arica, Azapa, Aug 1925, Werdermann 704".

Life cycle: Perennial. **Flowering period**: July to January. **Distribution**: Northern Chile, Region of Arica and Parinacota and Region of Tarapacá, up to 1340 m. **Conservation status**: Not evaluated. **References**: [22, 46, 47]. **Representative specimen**: Arica and Parinacota Region: Azapa, 50 m, Oehrens 34 (CONC).

20. ***Danthonia araucana*** Phil., Anales Univ. Chile 94: 31. 1896. **Typus**: Chile: In sylva Araucariarum montium Nahuelbuta, januario 1877 legimus, F. Philippi s.n. (HT: SGO).

Life cycle: Perennial. **Flowering period**: October to January. **Distribution**: This species grows in the regions of Valparaíso, Maule, Biobío, Arauco, Los Ríos and Los Lagos, from 10 to 930 m.a.s.l., in natural meadows, clay soils. It has been collected in *Eucalyptus* plantations. **Conservation status**: Not evaluated. **References**: [48]. **Representative specimen**: Biobío Region, Concepción, Chivilingo Norte, 180 m, Baeza & Rodríguez 2391 (CONC).

21. ***Danthonia chilensis*** E. Desv. var. ***aureofulva*** (E. Desv.) C.M. Baeza, Sendtnera 3: 32, f. 4i, 1996. **Basionym**: *Danthonia aureofulva* E. Desv., Hist. Fis. Pol. Chile, Bot. 6: 362. 1854. **Typus**: Chile, L'examplare unique a été remis à M. Gay pour servir de modèle au coloriste, Em. Desvaux, avril 1854 (HT: P).

Life cycle: Perennial. **Flowering period**: October to January. **Distribution**: This species grows in the regions of Valparaíso, O'Higgins, Biobío, Arauco, Los Ríos, and Los Lagos, from 5 to

630 m.a.s.l., in clay soils. It has been collected in the "espinal" of *Acacia caven*. **Conservation status**: Not evaluated. **References**: [12, 13, 48]. **Representative specimen**: Valparaíso Region: Viña del Mar, Fundo las Siete Hermanas, 280 m, Garaventa 3289 (CONC).

22. ***Danthonia malacantha*** (Steud.) Pilg., Notizbl. Bot. Gard. Berlin-Dahlem 10: 759. 1929. **Basionym**: *Trisetum malacanthum* Steud., Syn. Pl. Glumac. 1: 424. 1854. **Typus**: Chile: Huiti in pascuis, Jan 1852, Lechler 749 (P).

Life cycle: Perennial. **Flowering period**: October to March. **Distribution**: This species grows in north, central, and southern continental Chile as well as in the Juan Fernández Islands (Masatierra), in natural meadows and clay soils, usually below 1000 m.a.s.l. Regions of Coquimbo, Valparaíso, O'Higgins, Maule, Biobío, Arauco, Los Ríos, and Juan Fernández Archipelago (Masatierra). **Conservation status**: Not evaluated. **References**: [12, 13, 17, 22, 48]. **Representative specimen**: Region of Valparaíso, Juan Fernández, Masatierra, Stuessy et al. 11216 (CONC).

23. ***Deschampsia looseriana*** Parodi, Darwiniana 8: 460. 1949. **Typus**: Chile: Santiago, Batuco, 500 m, 17 Sep 1936, G. Looser 3439 (HT: BAA). **Synonym**: *Deschampsia looseriana* Parodi var. *triandra* Parodi, Darwiniana 8: 464. 1949.

Life cycle: Annual. **Flowering period**: September to December. **Distribution**: North-central Chile, in the regions of Coquimbo, Valparaíso, Metropolitan, O'Higgins, Maule and Biobío, from 10 to 2475 m. It has been collected in prairies and vernal pools. **Conservation status**: Not evaluated. **References**: [12, 13, 22, 49, 50]. **Comments**: Some authors [12] consider *D. looseriana* var. *triandra* as a valid species. This taxon differs from the typical variety by the presence of three stamens (one or two stamens in the typical form). However, flowers with one stamen were found also in the type of var. *triandra*, a character associated to cleistogamy [50]. **Representative specimen**: Coquimbo Region: Combarbalá, Cuesta de Punitaqui, 300 m, Marticorena & Matthei 387 (CONC).

24. ***Eragrostis kuschelii*** Skottsb., Ark. Bot. 4(15): 485. 1963. **Type**: Chile, Islas Desventuradas, San Ambrosio, Johow form 3, 5–6 Oct 1896 (UPS).

Life cycle: Perennial. **Distribution**: A local endemic found only in San Ambrosio Island (Valparaiso Region), from the sea level to 100 m of altitude. **Conservation status**: Not evaluated. **References**: [12, 13, 22, 51]. **Representative specimen**: Valparaiso Region, Isla San Ambrosio, Johow 2667 (SGO).

25. ***Eragrostis pycnantha*** (Phil.) Nicora, Gayana Bot. 51: 4. 1994. **Basionym**: *Poa pycnantha* Phil., Anales Univ. Chile 94: 165. 1896.

Life cycle: Perennial. **Distribution**: A very rare endemic with a distribution restricted to the Atacama Region at 140 m of altitude. **Conservation status**: Endangered (EN). **References**: [12, 13, 22, 51]. **Representative specimen**: Atacama Region, Huasco, Freirina, 140 m, Muñoz & Johnson 1982 (SGO).

26. ***Festuca panda*** Sw., J. Wash. Acad. Sci. 26(5): 209. 1936. **Type**: Coquimbo Region, Illapel, Cajón de Los Pelambre, alt. 2900 m, Jan 1932, Looser 2151 (US).

Life cycle: Perennial. **Flowering period**: February. **Distribution**: *Festuca panda* is a local endemic species, found only in the Coquimbo Region from 2900 to 3400 m, in Andean steppes and Atacama Desert. **Conservation status**: Not evaluated. **References**: [12, 13, 22, 52, 53]. **Representative specimen**: Coquimbo Region, Illapel, Hacienda Cuncumén, Cajón de Los Pelambres, 2900 m, Looser 2151 (isotype, CONC).

27. ***Hierochloe altissima*** Steud., Syn. Pl. Glumac. 1: 13. 1854. **Type**: "Lechler legit Valdiviae".

Life cycle: Perennial. **Distribution**: Southern Chile, it has been collected in the Regions of Biobío, Araucanía, Los Ríos, Los Lagos, and Aysen, from 5 to 780 m.a.s.l. **Conservation status**: Not evaluated. **References**: [12, 13, 22, 32, 54–56]. **Comments**: Genus *Hierochloe* was considered a synonym of genus *Anthoxanthum* [57]. However, based on evidence from floral biology, it was suggested that *Anthoxanthum* and *Hierochloe* should be maintained as different genera [58]. **Representative specimen**: Biobío Region, Concepción, Soreng & Soreng 7034 (CONC).

28. ***Hierochloe spicata*** Parodi, Revista Mus. La Plata 3(14): 196. 1941. **Type**: "Chile, Magallanes, Philippi s.n."

Life cycle: Perennial. **Distribution**: Regions of Araucanía, Los Ríos, and Magallanes y Antártica Chilena. **Conservation status**: Not evaluated. **References**: [12, 13, 22, 54–56]. **Representative specimen**: Araucanía Region, Malleco, Lonquimay, Montero 4810 (CONC).

29. ***Hordeum brachyatherum*** Phil., Anales Univ. Chile 94: 346. 1896. **Type**: Chile, habitat in Andibus de Linares dicti, Ortega s.n. (SGO).

Life cycle: Perennial. **Distribution**: Regions of Coquimbo, Valparaíso, Maule, and Magallanes, from 2 to 500 m.a.s.l. **Conservation status**: Not evaluated. **References**: [12, 13, 22]. **Comments**: Although this species was cited for the Patagonia of Argentina (Río Negro and Chubut) [59], probably based on misidentified specimens [60]. **Representative specimen**: Valparaíso Region, Petorca, Pichicuy, Marticorena *et al.* 161 (CONC).

30. ***Imperata parodii*** Acevedo, Bol. Soc. Argent. Bot. 12: 358. 1968. **Type**: Chile, Prov. Cautín, al S de la boca del Toltén. Reiche s.n. (SGO).

Life cycle: Perennial. **Distribution**: A local, rare endemic, known only from the type specimen collected in the Araucanía Region. **Conservation status**: Not evaluated. **References**: [12, 13, 22, 61]. **Representative specimen:** Araucanía Region, Cautín, S de la boca del Toltén, Reiche s.n. (SGO, type).

31. ***Jarava tortuosa*** (E. Desv.) Peñailillo, Gayana, Bot. 59(1): 32. 2002. **Basionym**: *Stipa tortuosa* E. Desv., Hist. Fis. Pol. Chile, Bot. 6: 281. 1854. **Type**: Cordillera de Doña Ana, Gay s.n. (HT: P).

Life cycle: Perennial. **Flowering period**: September to January. **Distribution**: This species is endemic to northern Chile. It grows in the regions of Antofagasta, Atacama, and Coquimbo, from the sea level to 900 m, mainly on the coast. **Conservation status**: Not evaluated. **References**: [12, 13, 22, 62]. **Representative specimen**: Antofagasta Region: Taltal, Quebrada Changos, 250 m, Ricardi 2575 (CONC).

32. ***Megalachne berteroniana*** Steud., Syn. Pl. Glumac. 1: 237. 1854. **Type**: Chile, Juan Fernández Islands, Jan–Jun 1830, Bertero 1440 (P).

Life cycle: Perennial. **Distribution**: Endemic to the Juan Fernández Archipelago (Masatierra Island), 180–670 m. **Conservation status**: Vulnerable (VU). **References**: [12, 13, 22, 63]. **Representative specimen**: Masatierra Island, Valle Villagra, Stuessy *et al.* 6521 (CONC).

33. ***Megalachne masafuerana*** (Skottsb. & Pilg.) Matthei, Bol. Soc. Biol. Concepción 48: 171. 1974. **Basionym**: *Bromus masafueranus* Skottsb. & Pilg., Repert. Spec. Nov. Regni Veg. 16: 385. 1920. **Type**: Chile: Juan Fernández, Masafuera, Hochland bei Las Torres, sterile Felsen, 1370 m, C. & I. Skottsberg 145 (GB).

Life cycle: Perennial. **Distribution**: Endemic to the Juan Fernández Archipelago (Masafuera and Masatierra islands), 320–1100 m. **Conservation status**: Critically endangered (CR). **References**: [12, 13, 22, 63]. **Representative specimen**: Juan Fernández Islands, Masafuera, Quebrada Angosta, Meyer 9405. CONC).

34. ***Melica argentata*** E. Desv., Hist. Fís. Pol. Chile, Bot. 6: 374. 1854. **Type**: Chile, Rancagua.

Life cycle: Perennial. **Flowering period**: September to February. **Distribution**: Regions of Antofagasta, Coquimbo, Valparaíso, Metropolitan, O'Higgins, and Maule from 15 to 2200 m in the Cordillera de la Costa, longitudinal valley and Cordillera de Los Andes. **Conservation status**: Not evaluated**.** **Reference**: [12, 13, 22, 64]. **Representative specimen**: Coquimbo Region: Choapa, Cuncumén, 1400 m, Jiles 4292 (CONC).

35. ***Melica commersonnii*** Nees ex Steud., Syn. Pl. Glumac. 1: 290. 1854. **Type**: Chile?

Life cycle: Perennial. **Flowering period**: August to December. **Conservation status**: Not evaluated**.** **References**: [12, 13, 22, 64]. **Distribution**: From the sea level to 1900 m of elevation in the Regions of Coquimbo, Valparaíso, Metropolitan, O'Higgins, Maule, and Biobío. **Representative specimen**: Biobío Region: Ñuble, camino de Pullay a Quile, 275 m, Baeza & López 2247 (CONC).

36. ***Melica longiflora*** Steud., Syn. Pl. Glumac. 1: 290. 1854. **Type**: Chile, Bertero hb. 1816 "*M. laxiflora* ". **Common name**: Lengua de gato.

Life cycle: Perennial. **Flowering period**: August to December. **Distribution**: Regions of Coquimbo, Valparaíso, Metropolitan, O'Higgins, and Maule, from 20 to 1300 m.a.s.l. **Conservation status**: Not evaluated. **References**: [12, 13, 22, 64]. **Representative specimen**: Coquimbo Region: Ovalle, Estancia Talca, 300 m, Jiles 446 (CONC).

37. ***Melica mollis*** Phil., Anales Univ. Chile 94: 161. 1896. Type: Chile: prope Carrizal, G. Geisse s.n. (SGO).

Life cycle: Perennial. **Flowering period**: November to February. **Distribution**: Endemic to northern Chile, it has been collected in the regions of Atacama and Coquimbo, 340 m.a.s.l. **Conservation status**: Not evaluated**.** **References**: [12, 13, 22, 64]. **Representative specimen**: Atacama Region: Carrizal Bajo, Mina Oriente, Muñoz-Schick 3055 (SGO).

38. ***Melica paulsenii*** Phil., Anales Univ. Chile 94: 159. 1896. **Type**: Chile, in praedio S. Isidro prope Quillota, F. Philippi s.n. (SGO).

Life cycle: Perennial. **Flowering period**: September to December. **Distribution**: Regions of Coquimbo, Metropolitan, Valparaíso, and Maule between 50 and 700 m.a. s.l. **Conservation status**: Not evaluated. **References**: [12, 13, 22, 64]. **Representative specimen**: Maule Region: Curicó, Laguna Vichuquén, 20 m, Zöllner 1954 (CONC)

39. ***Melica poecilantha*** E. Desv., Hist. Fís. Pol. Chile, Bot. 6: 379. 1854. **Type**: [Chile], crece por copitas entre los peñascos cubiertos de arbustos, en La Serena y Arqueros, Prov. Coquimbo, Gay s.n.

Life cycle: Perennial. **Flowering period**: October to November. **Distribution**: Endemic to the Coquimbo Region from 500 to 1200 m. **Conservation status**: Near threatened (NT). **References**: [12, 13, 22, 64]. **Representative specimen**: Coquimbo Region: Illapel, 13 km N de Mantos de Hornillo, Quebrada Pajaritos, 280 m, Marticorena *et al.* 367 (CONC).

40. ***Melica violacea*** Cav., Icon. 5: 472. 1799. **Type**: "Habitat prope Talcahuano in Chile, floretque Februario et Martio." **Common name**: Pasto bandera.

Life cycle: Perennial. **Flowering period:** September to February. **Distribution:** Regions of Coquimbo, Valparaíso, Metropolitan, O'Higgins, Maule, Biobío, Araucanía, and Los Ríos, between the sea level and 2000 m.a.s.l., in coastal zones, longitudinal valley, and Cordillera de Los Andes at low elevations. It grows in sandy and rocky soils, and under sub-Andean forest; it has been collectively associated to *Kageneckia* sp., *Peumus boldus*, *Elythraea* sp., *Colletia* sp., *Bomarea* sp. and *Tristerix* sp. and *Muhlenbeckia* sp. **Conservation status**: Not evaluated. **References**: [12, 13, 22, 64]. **Representative specimen:** Biobío Region: Ñuble, General Cruz, Arenales del Itata, 101 m, Montero 1931 (CONC).

41. ***Nassella barbinodis*** (Phil.) M. Muñoz, Ciald. & Morrone, Darwiniana n.s. 1(1): 90. 2013. **Basionym**: *Stipa barbinodis* Phil., Anales Univ. Chile 93: 721. 1896. **Type**: Chile. XIV Región: Valdivia, fundo de San Juan, I-1887, R. A. Philippi s.n.

Life cycle: Perennial. **Flowering period**: November to January. **Distribution**: Southern Chile. This species is found in the regions of Biobío, Araucanía, Los Ríos, and Los Lagos from 60 to 280 m.a.s.l. It has been collected on sandy soil, in bushes with grasses, and *Juncus* sp. **Conservation status**: Not evaluated. **References**: [12, 13, 22, 29, 65]. **Comments**: Matthei [29] treated *Stipa barbinodis* Phil. as a synonym of *S. duriuscula* Phil. Other authors [12, 18] accept this treatment and included *Nassella barbinodis* in the synonym of *Nassella duriuscula* (Phil.) Barkworth. Later, Cialdella et al. [65] based on several differences of the spikelets consider them as two different species. The types in US of both species differ in the characteristics of the floret (floret with crown inconspicuous, 5 mm long in *N. duriuscula*; floret with a conspicuous crown, 8 mm long in *N. barbinodis*). **Representative specimen**: Biobío Region, Concepción, Hualpén, Parque Pedro del Río, 150 m, Gunckel 10068 (CONC).

42. ***Nassella chilensis*** (Trin.) E. Desv. var. ***juncea*** (Phil.) M. Muñoz, Gayana, Bot. 47: 22. 1990. **Basionym**: *Nassella juncea* Phil., Linnaea 33(3–4): 277. 1864. **Type**: Chile, prope Corral detexit ornat. H. Krause s.n.

Life cycle: Perennial. **Flowering period**: November to February. **Distribution**: This variety grows in south-central Chile, in the Metropolitan, Biobío, Araucanía, and Los Ríos regions, from 20 to 925 m of elevation. **Conservation status**: Not evaluated. **References**: [12, 13, 22, 65–67]. **Comments**: As Cialdella et al. pointed out [65], Barkworth accepted this taxon as species [67]. In this paper, we follow the treatment proposed by Cialdella et al. [65]. **Representative specimen**: Metropolitan Region: Chacabuco, Altos de Chicauma, desde tranque hacia Morro Jarillas, 2100 m, García 3762 (CONC).

43. ***Nassella coquimbensis*** (Matthei) Peñail., Gayana, Bot. 55(2): 86. 1998[1999]. **Basionym**: *Stipa coquimbensis* Matthei, Gayana, Bot. 13: 35. 1965. **Type**: Chile: IV Región: Prov. Coquimbo, Carretera Panamericana, 8 km al norte de la Quebrada del Teniente, 13-X-1963, Marticorena & Matthei 163 (CONC).

Life cycle: Perennial. **Flowering period**: October. **Distribution**: Endemic to the Region of Coquimbo, between 60 and 350 m of elevation, in clay-rocky soils. **Conservation status**: Not evaluated. **References**: [12, 13, 22, 29, 65]. **Representative specimen**: Coquimbo Region: Illapel, Huentelauquén, 60 m, Jiles 5754 (CONC).

44. ***Nassella duriuscula*** (Phil.) Barkworth, Taxon 39(4): 610. 1990. **Basionym**: *Stipa duriuscula* Phil., Linnaea 33(3–4): 282. 1864. **Type**: Chile, VI Región de O'Higgins, Prov. Colchagua, XI-1860, L. Landbeck s.n.

Life cycle: Perennial. **Flowering period**: November to December. **Distribution**: Central Chile, in the regions of Coquimbo, Metropolitan, O'Higgins, and Maule, between 15 and 1000 m.a.s.l. **Conservation status**: Not evaluated. **References**: [12, 13, 22, 29, 65, 67]. **Representative specimen**: Coquimbo Region: Illapel, Cavilolén, Jiles 2670 (CONC).

45. ***Nassella gibba*** (Phil.) Muñoz-Schick, Gayana, Bot. 47: 26. 1990. **Basionym:** *Piptochaetium gibbum* Phil., Anales Univ. Chile 93: 731. 1896. **Type**: Chile. VII Región del Maule: Talca, R. Iturriaga s.n.

Life cycle: Perennial. **Flowering period**: October to January. **Distribution**: Coast and longitudinal valley of the regions of Valparaíso, Metropolitan, O'Higgins, Maule, and Biobío, from 15 to 1000 m of elevation. **Conservation status**: Not evaluated. **References**: [12, 13, 22, 65, 66]. **Representative specimen**: Maule Region: Talca, R.N. Empedrado, El Fin, 300 m, Finot & López 1335 (CONC).

46. ***Nassella hirtifolia*** (Hitchc.) Barkworth, Taxon 39(4): 610. 1990. **Basionym**: *Stipa hirtifolia* Hitchc., Contr. U. S. Natl. Herb. 24(7): 285. 1925. **Type**: Chile.

Life cycle: Perennial. **Flowering period**: September to January. **Distribution**: Regions of Coquimbo, Valparaíso, Metropolitan, and O'Higgins, from 20 to 1460 m.a.s.l. **Conservation status**: Not evaluated. **References**: [12, 13, 22, 29, 65]. **Comments:** *Nassella hirtifolia* is close to *N. laevissima*, which differs from *N. hirtifolia* having florets totally glabrous (glabrous or pilose on the lower two-third in *N. hirtifolia*), antopodium 0.2–0.3 mm (0.5–1.0 mm long in *N. hirtifolia*) with hairs shorter than half the length of the lemma (hairs surpassing the lower one-third of the length of the lemma in *N. hirtifolia*). **Representative specimen**: Metropolitan Region: Río Clarillo, 800 m, Araya s.n. (CONC).

47. ***Nassella lachnophylla*** (Trin.) Barkworth, Taxon 39(4): 610. 1990. **Basionym:** *Stipa lachnophylla* Trin., Mem. Acad. Imp. Sci. Saint-Petersbourg, Ser. 6, Sci. Math., Seconde Pt. Sci. Nat. 4,2(1): 39. 1836. **Type**: Chile, 1832, J. D. Prescott & H. Cuming s.n.

Life cycle: Perennial. **Flowering period**: October to December. **Distribution**: Regions of Coquimbo, Valparaíso, Metropolitan, O'Higgins, Maule, and Araucanía between 10 and 1000 m of elevation; it has been collected in plowed fields, edge of roads, and ponds. **Conservation status**: Not evaluated. **References**: [12, 13, 22, 29 65]. **Representative specimen**: Metropolitan Region: Santiago, Cerro San Cristóbal, 800 m, Mahu 4084 (CONC).

48. ***Nassella macrathera*** (Phil.) Barkworth, Taxon 39(4): 610. 1990. **Basionym:** *Stipa macrathera* Phil., Anales Univ. Chile 93: 720. 1896. **Type**: Chile, Santiago, in collibus de Renca, R. A. Philippi s.n.

Life cycle: Perennial. **Flowering period**: September to February. **Distribution**: Regions of Coquimbo, Metropolitan, Maule, Biobío, Arauco, and Los Ríos from 2 to 2000 m. It grows also in Juan Fernández Archipelago [65]. **Conservation status**: Not evaluated. **References**: [12, 13, 22, 29, 65]. **Representative specimen**: Araucanía Region: Cautín, Temuco, 110 m, Gunckel 11021 (CONC).

49. ***Nassella parodii*** (Matthei) Barkworth, Taxon 39(4): 611. 1990. **Basionym**: *Stipa parodii* Matthei, Gayana Bot. 13: 89. 1965. **Type**: Chile, VIII Región del Bio Bío, Prov. Ñuble, 3 km pasado de San Nicolás, camino a Quirihue, 5-XI-1961, O. Matthei 245 (Holotype, CONC 52424).

Life cycle: Perennial. **Flowering period**: November to January. **Distribution**: Central Chile, in the regions of O'Higgins and Biobío between 5 and 370 m.a.s.l. It grows in the *Acacia caven* "espinal" of the Secano interior of the Province of Ñuble and also on the shores of the sea on brackish soils. **Conservation status**: Not evaluated. **References**: [12, 13, 22, 29, 65]. **Comments**: This species is morphologically similar to *N. macrathera* from which it differs by the following characteristics: lemma papillose only toward the apex (lemma minutely papillose over the entire surface in *N. macrathera*), floret 1.1–1.4 cm long (floret 0.9–1.2 mm long in *N. macrathera*), lemma glabrous (lemma hairy on the lower part of the dorsal nerve in *N. macrathera*). **Representative specimen**: O'Higgins Region: Cardenal Caro, 19 km E of Pichilemu, 2.6 km W of Puente Los Valles 1 on Hwy to Nancagua, 370 m, Lammers *et al.* 7885 (CONC).

50. ***Nassella pfisteri*** (Matthei) Barkworth, Taxon 39(4): 611. 1990. **Basionym**: *Stipa pfisteri* Matthei, Gayana Botánica 13: 100. 1965. **Type**: Provincia de Maule, Camino de Parral a Cauquenes km 36, 11-I-1964, Marticorena & Matthei 488 (CONC).

Life cycle: Perennial. **Flowering period**: September to January. **Distribution**: Endemic to South-Central Chile, where it is found in the regions of Maule and Biobío, between 60 and 360 m of elevation. **Conservation status**: Not evaluated. **References**: [12, 13, 22, 29, 65]. **Comments**: Cialdella et al. [65] cited the specimen Matthei 259 as type of *S. pfisteri*. **Representative specimen**: Biobío Region, Ñuble, camino a San Nicolás, 100 m, Matthei 252 (CONC).

51. ***Nassella pungens*** E. Desv., Hist. Fis. Pol. Chile, Bot. 6: 268. 1854. **Type**: Chile, San Fernando in collibus, Gay 69.

Life cycle: Perennial. **Flowering period**: September to May. **Distribution**: North-Central Chile: Regions of Arica and Parinacota, Tarapacá, Antofagasta, Atacama, Coquimbo, and Metropolitan, between 5 and 1000 m.a.s.l. **Conservation status**: Not evaluated. **References**: [12, 13, 22, 65, 66]. **Representative specimen**: **Arica y Parinacota Region**: East of Arica at Termas Jurasi above Putre, Peterson & Soreng 15742 (CONC).

52. ***Paspalum chilense*** Catanzaro & G.H. Rua, Phytotaxa 197(4): 247. 2015. **Type**: Chile, Región de Los Ríos, provincia de Ranco, Lago Ranco, río Calcurrupe, 40°13′32.8″S, 72°14′22.4″W, 81 m, 18 December 2007, G.H. Rua, M. Medina-Nicolas & E. Piel 918 (BAA).

Life cycle: Perennial. **Flowering period**: December. **Distribution**: This species was described recently [68] based on material collected in the regions of Biobío, Araucanía, and Los Ríos. **Conservation status**: Endangered (EN) [68]. **References**: [68]. **Reference specimen**: Los Ríos Region, Lago Ranco, río Calcurrupe, 81 m, G.H. Rua *et al.* 918 (BAA, type).

53. ***Paspalum forsterianum*** Fluggé, Gram. Monogr., Paspalum 165. 1810. Type: Nova Caledonia, Forster s.n. **Common name**: Mauku toa, heriki hare.

Life cycle: Perennial. **Distribution:** This species is endemic to Eastern Island where it grows on sea cliffs of Rapa Nui and Moto Nui. **Conservation status**: Not evaluated. **References**: [6, 12, 13, 22]. **Representative specimen**: Isla de Pascua, Costa de Motu Tautara, Zizka 593 (CONC).

54. ***Phalaris amethystina*** Trin., Mém. Acad. Imp. Sci. Saint-Pétersbourg, Sér. 6, Sci. Math., Seconde Pt. Sci. Nat. 5,3(3): 56. 1839. **Type**: Chile, in pascuis herbidis aquosis montis La Leona Rancagua, Bertero 534 (Lecotype: LE; Isolectotype: SGO!); lectotype designated by Hitchcock in Jepson, Fl. Calif. 3: 96–99.

Life cycle: Annual. **Flowering period**: September to January. **Distribution**: Regions of Coquimbo, Valparaíso, Metropolitan, O'Higgins, Maule, Biobío, Araucanía, and Los Lagos. It is also found in Juan Fernández Archipelago (Masatierra) [9, 16, 69]. It grows mainly at low elevations up to 1000 m.a.s.l., associated to the "espinal" of *Acacia caven* (Mol.) Mol., in rich, loamy soils with shallow water. **Conservation status**: Not evaluated. **References**: [9, 12, 13, 22, 70–73].**Representative specimen**: Coquimbo Region, Coquimbo, Ovalle, Quebrada Toigoncillo, 250 m, Jiles 852 (CONC).

55. ***Piptochaetium angolense*** Phil., Anales Univ. Chile 93: 734. 1896. **Type:** Chile. "Angol, Novemb. 1887", Philippi s.n. (HT: SGO).

Life cycle: Perennial. **Distribution**: Regions of Coquimbo, Valparaíso, Metropolitan, O'Higgins, Maule, Biobío, and Araucanía from 10 to 740 m.a.s.l. **Conservation status**: Not evaluated. **References**: [12, 13, 22, 74]. **Representative specimen**: Valparaíso Region: Petorca, Pichicuy, Marticorena *et al.* 189 (CONC);

56. ***Poa acinaciphylla*** E. Desv., Hist. Fis. Pol. Chile, Bot. 6: 412. 1854. Type: Chile, Gay 1119, Cat. Propr. Synomym: *Poa villaroelii* Phil., Anales Univ. Chile 94: 169. 1896.

Life cycle: Perennial. Flowering period: January to February. Distribution: North-central Chile, in the regions of Coquimbo, Valparaíso, and Metropolitan. Conservation status: Not evaluated.

References: [12, 13, 22, 75]. Representative specimen: Metropolitan Region: Santiago, Valle Nevado, Negritto 961 (CONC).

57. ***Poa pfisteri*** Soreng, J. Bot. Res. Inst. Texas 2(2): 850. 2008. **Type**: Chile, Region VIII: Bio-BÍo, Province of Santa Bárbara, Puente Mininco, 1 Nov 1943, A. Pfister s.n.

Life cycle: Perennial. **Flowering period**: November. **Distribution**: A rare endemic known only from the type collection from Santa Bárbara [Province of Biobío], between 200 and 300 m. **Conservation status**: Not evaluated. **References**: [13, 75]. **Representative specimen**: Biobío Region, Biobío Prov., Santa Bárbara, Puente Mininco, Pfister s.n. (US, CONC, type).

58. ***Poa cumingii*** Trin., Mem. Acad. Imp. Sci. Saint.Petersbourg, Sér. 6. Sci. Math., Seconde Pt. Sci. Nat. 4,2(1): 66. 1836.

Life cycle: Perennial. **Flowering period**: November to February. **Distribution**: A fairly common species distributed from Coquimbo to Chiloé (Regions of Coquimbo, Valparaíso, Maule, Biobío, Araucanía, Los Ríos, and Los Lagos) between 3 and 350 m. **Conservation status**: Not evaluated. **References**: [13, 75]. **Representative specimens**: Biobío Region, Concepción, Playa de Lenga, 10 m, L. Giussani s.n. (CONC).

59. ***Poa gayana*** E. Desv., Hist. Fis. Pol. Chile, Bot. 6: 416. 1854. **Type**: Chile: Cordilleras de Chile, Gay s.n.

Life cycle: Perennial. **Flowering period**: October to April. **Distribution**: Regions of Coquimbo, Valparaíso, Metropolitan, O'Higgins, Maule, and Biobío between 470 and 3400 m of elevation. **Conservation status**: Not evaluated. **References**: [12, 13, 18, 22, 75]. **Comments**: Some authors consider that this taxon includes *Poa lanuginosa* var. *neuquina;* under this concept, *P. gayana* grows in Chile and Argentina [18]; this idea was not accepted in recent treatments of *Poa* for Argentina [76]. The taxonomy of *Poa gayana* is difficult and its status needs to be clarified [75]. **Representative specimen**: Maule Region: Talca, Central Los Cipreses, Quebrada El Ciego, Finot & López 2011 (CONC).

60. ***Poa paposana*** Phil., Fl. Atacam. 55: 1860. Type: Chile, Antofagasta Region: In regione herbosa prope Paposo inveni.

Life cycle: Perennial. **Flowering period**: August to November. **Distribution**: North-central Chile: Regions of Antofagasta, Atacama, Coquimbo, Valparaíso, and Metropolitan, between 30 and 820 m. **Conservation status**: Not evaluated. **References**: [12, 13, 75]. **Representative specimen**: Coquimbo Region, Limarí, Parque Nacional Fray Jorge, Soreng & Soreng 7055 (CONC).

61. ***Poa schoenoides*** Phil., Anales Univ. Chile 94: 166. 1896. **Type**: Chile: Prope Queñi in Andibus prov. Valdivia, O. Philippi s.n.

Life cycle: Perennial. **Distribution**: Southern Chile, Los Ríos Region. **Conservation status**: Not evaluated. **References**: [12, 13, 18, 22]. **Comments**. The taxonomy of this species needs to be clarified [13].

62. ***Poa trachyantha*** Hack., Repert. Spec. Nov. Regni Veg. 10(243–247): 173. 1911. **Type:** In expeditione ad flumen Aysen, P. Dusén 568 (W).

Life cycle: Perennial. **Distribution**: Aysen Region. **Conservation status**: Not evaluated. **References**: [12, 13, 18, 22]. **Representative specimen:** Aysen Region, Río Aysen, Dusén 568 (US).

63. ***Poa tricolor*** Nees ex Steud., Syn. Pl. Glumac. 1: 259. 1854. Type: Chile, Valparaíso, Cumming s.n. (B).

Life cycle: Perennial. **Distribution**: Valparaíso Region. **Conservation status**: Not evaluated. **References**: [12, 13, 18, 22]. **Representative specimen**: Chile, Valparaíso Region, Cuming 468 (US, type).

64. ***Podophorus bromoides*** Phil., Anales Univ. Chile 13: 169. 1856. **Typus**: CHILE: frequens in insula Juan Fernández, Germain s.n. (LT: SGO 37130).

Life cycle: Probably perennial. **Flowering period**: October. **Distribution**: Restricted to the Robinson Crusoe Island (Masatierra or Más a Tierra), Juan Fernández Archipelago. **Conservation status**: Extinct (EX). **References**: [12, 13, 16, 18, 22, 77, 78, 79]. **Comments**: After the collection made by German in 1854, this species has not been collected again. Skottsberg [17] referred the genera *Brachyelytrum* and *Aphanelytrum* as the nearest relatives of *Podophorus*. **Representative specimen**: Juan Fernández Island (Masatierra), Germain s.n. (SGO, type).

65. ***Polypogon cachinalensis*** Phil., Fl. Atacam. 54. 1860. **Type**: Chile: Ad Cachinal de la Sierra, 25° lat. m., 7000 p.s.m. legi, Philippi 341 (SGO).

Life cycle: Perennial. **Distribution**: Endemic to northern Chile; it has been collected in the regions of Antofagasta and Coquimbo between 2600 and 3000 m. **Conservation status**: Not evaluated. **References**: [12, 13, 80]. **Representative specimen**: Coquimbo Region: Cordillera de Combarbalá, Potrero Grande, 2700–2800 m, Jiles 4824 (CONC).

66. ***Polypogon elongatus*** Kunth var. ***strictus*** E. Desv., Hist. Fis. Pol. Chile, Bot. 6: 302. 1854. **Type**: Chile: Santiago, Gay s.n. (SGO).

Life cycle: Perennial. **Flowering period**: December to January. **Distribution**: It has been collected in the Regions Metropolitan, Los Ríos, and Los Lagos between 5 and 200 m.a.s.l. **Conservation status**: Not evaluated. **References**: [12, 13, 22, 80]. **Representative specimen**: Los Ríos Region: Valdivia, Corral, La Rama, 6 m, Gunckel 40920 (CONC).

67. ***Polypogon linearis*** Trin., Linnaea 10(3): 301. 1841. Type: Chile bor. in campis ad Concón.

Life cycle: Annual. **Flowering period**: September to February. **Distribution**: Regions of Coquimbo, Valparaíso, Metropolitan, O'Higgins, Maule, and Biobío, from 20 to 740 m. **Conservation status**: Not evaluated. **References**: [12, 13, 22, 80]. **Representative specimen**: Metropolitan Region, Santiago, Peñaflor, Montero 193 (CONC).

68. ***Rytidosperma paschale*** (Pilger) C.M. Baeza, Gayana Botánica 47(3–4): 84. 1990. **Basionym**: *Danthonia paschalis* Pilger, *in* Skottsberg, Nat. Hist. Juan Fernández 2: 67. 1922. **Type**: Chile. Isla de Pascua, on the slope of Mountain Katiki, 16-VI-1917, Skottsberg & Skottsberg 658.

Life cycle: Perennial. **Flowering period**: January to June. **Distribution**: Endemic to Easter Island; it was collected in mount Katiki and Rano Kao volcano [4]. **Conservation status**: Not evaluated. **References:** [4, 12, 13, 48]. **Representative specimen**: Rano Kao, interior del cráter, 200 m, Etienne s.n. (CONC).

69. ***Rytidosperma quirihuense*** C. Baeza, Novon 12: 31. 2002. **Type**: Chile. Ñuble: Camino Las Achiras hacia Quirihue (36°13′S, 72°45′W), 355 m, 20 Nov. 2000, C. Baeza, P. López & M. Parra 2112 (HT: CONC).

Life cycle: Perennial. **Distribution**: *Rytidosperma quirihuense* is found in Valparaíso Region and Biobío Region from 355 to 390 m.a.s.l. **Conservation status**: Not evaluated. **References**: [16]. **Ilustrations**: [16]. **Representative specimen**: Biobío Region, Ñuble, camino Las Achiras hacia Quirihue, 355 m, Baeza *et al.* 2112 (CONC, holotype).

70. ***Trisetum johnstonii*** (Louis-Marie) Finot subsp. ***mattheii*** (Finot) Finot, Curr. Topics Plant Biol. 11: 61. 2010. **Basionym**: *Trisetum matthei* Finot, Ann. Missouri Bot. Gard. 92(4): 551. 2005. **Type**: Chile, Región I: Tarapacá, camino de Arica al Portezuelo de Chapiquiña, km 111, 10°18′S, 69°30′W, 4100 m, C. Marticorena *et al.* 86 (CONC).

Life cycle: Perennial. **Flowering period**: February. **Distribution**: This species is restricted to Arica and Parinacota Region, 4100 m. This subspecies is known only from the type collection. **Conservation status**: Not evaluated. **References**: [12, 13, 81]. **Representative specimen**: Tarapacá Region, camino de Arica al Portezuelo de Chapiquiña, Marticorena *et al.* (CONC, type).

71. ***Trisetum nancaguense*** Finot, Ann. Missouri Bot. Gard. 92(4): 553. 2005. **Type**: Chile, Región VI: Prov. Cardenal Caro, 12 km E of Pichilemu on Hwy toward Nancagua, T. Lammers *et al.* 7894 (CONC).

Life cycle: Perennial. **Flowering period**: November to February. **Distribution**: Regions Metropolitan, O'Higgins, Maule, and Biobío between 45 and 2450 m. **Conservation status**: Not evaluated. **References**: [12, 13, 81]. **Representative specimen**: O'Higgins Region, Pichilemu, Nancagua, Lammers et al. 7894 (CONC, type).

Acknowledgements

The authors wish to thank the project "Updated systematic list of the native vascular flora of Chile, origin and geographical distribution." Native Forest Research Fund Project, National Forestry Corporation (CONAF) No. 138/2016. We are grateful to the curators and technical staff of the herbaria of the University of Concepcion (CONC) and the National Museum of Natural History, Santiago (SGO).

Author details

Víctor L. Finot[1]*, Alicia Marticorena[2], Roberto Rodríguez[2] and Romina G. Muñoz[1]

*Address all correspondence to: vifinot@udec.cl

1 Department of Animal Production, Faculty of Agronomy, University of Concepción, Chillán, Chile

2 Department of Botany, Faculty of Natural Sciences and Oceanography, University of Concepción, Concepción, Chile

References

[1] Meiners Ochora M, Hernández López L. Únicamente en México: Especies endémicas y las plantas de Jalisco. Biodiversitas. 2007;**71**:10-15

[2] Chile, Ministerio del Medio Ambiente. Inventario nacional de especies endémicas de Chile. 2017. Available from: http://especies.mma.gob.cl/CNMWeb/Web/WebCiudadana/pagina.aspx?id=89. [Accessed: 28-02-2017]

[3] Scarbek C. A review of endemic species in the Eastern Arc Afromontane region: Importance, inferences, conservation. Macalester Reviews in Biogeography. 2009;**1**:1-21

[4] Finot VL, Barrera JA, Marticorena C, Rojas G. Systematic diversity of the family *Poaceae* (Gramineae) in Chile. In: Grillo O, Venora G, editors. The Dynamical Processes of Biodiversity-Case Studies of Evolution and Spatial Distribution. Croatia: Intech; 2011. pp. 71-108. DOI: 10.5772/23346. ch4

[5] Finot VL, Marticorena C, Barrera JA, Wilckens R, Fisher S, Silva F, Rojas G. Analysis of the diversity of the family *Poaceae* in the Region of Aysen, Chilean Patagonia. Current Topics in Plant Biology. 2012;**13**:57-70

[6] Finot VL, Marticorena C, Marticorena A, Rojas G, Barrera JA. Grasses (*Poaceae*) of Easter Island: Native and introduced species diversity. In: Blanco JA, Lo YH, editors. Biodiversity in Ecosystems: Linking Structure and Function. Croatia: Intech; 2015. pp. 383-406. DOI: 10.5772/59154. ch15

[7] Finot VL, Marticorena C, Barrera JA, Muñoz-Schick M, Negritto MA. Diversidad de la familia *Poaceae* (Gramineae) en la Región del Bio-Bío, Chile, basada en colecciones de herbario. Gayana Botanica. 2009;**66**:134-157

[8] Finot VL. Analysis of species boundaries in *Trisetum* Pers. Sect. *Trisetaera* Asch. & Graebn. (*Poaceae*) in America using statistical multivariate methods. Current Topics in Plant Biology. 2010;**11**:39-74

[9] Baeza CM, Marticorena C, Stuessy T, Ruiz M, Negritto MA. *Poaceae* en el Archipiélago de Juan Fernández (Robinson Crusoe). Gayana Botanica. 2007;**64**:125-174

[10] Rúgolo ZE, Molina AM. Las especies del género *Agrostis* L. (*Gramineae*: *Agrostideae*) de Chile. Gayana Botanica. 1997;**54**:91-156

[11] Muñoz-Schick M. Actualización de la flora del Parque Nacional Puyehue, Región del Los Lagos, Chile. Cloris Chilensis 17(2) [Internet]. 2014. Available from: www.clorischile.cl [Accessed: 28-02-2017]

[12] Zuloaga FO, Morrone O, Belgrano M. Catálogo de las plantas vasculares del Cono Sur [Internet]. 2009 [Updated: 2016]. Available from: http://www2.darwin.edu.ar/Proyectos/FloraArgentina/FA.asp [Accessed: 28-02-2017]

[13] Missouri Botanical Garden. Tropicos [Internet]. [Updated: 2017]. Available from: www.tropicos.org [Accessed: 28-02-2017]

[14] Moreno R, Le Quesne C, Díaz I, Rodríguez R. Vascular flora of Futangue Park, Región de Los Ríos (Chile). Gayana Botanica. 2013;**70**:121-135

[15] Marticorena C, Stuessy T, Baeza C. Catalogue of the vascular flora of Juan Fernández islands, Chile. Gayana Botanica. 1998;**55**:187-211

[16] Baeza CM, Stuessy T, Marticorena C. Notes on the *Poaceae* of the Robinson Crusoe (Juan Fernández) Islands. Brittonia. 2002;**54**:154-163

[17] Skottsberg C. The phanerogams of the Juan Fernández Islands. The Natural History of Juan Fernández and Easter Islands. Vol. 2. 1921. pp. 95-240

[18] Soreng RJ, Peterson PM, Davidse G, Judziewicz EJ, Zuloaga FO, Filgueiras TS, Morrone O. Catalogue of New World Grasses (*Poaceae*): IV. subfamily Pooideae. Contributions from the United States National Herbarium. 2003;**48**:1-730

[19] McCloskie G. Revision of Flora Patagonica. Reports of the Princeton Expeditions to Patagonia. 8, Bot., Suppl. 1915;**1**:1-307

[20] Spegazzini C. Nova addenda af floram Patagonicam. Partes III et IV. Anales del Museo Nacional De Buenos Aires. 1902;**7**:135-308

[21] Navas LE, editor. Flora de la Cuenca de Santiago de Chile. In: Andrés Bello, editor. Vol. 1. Santiago: Universidad de Chile; 1973. p. 301

[22] Marticorena C, Quezada M. Catálogo de la flora vascular de Chile. Gayana Botanica. 1985;**42**:1-157

[23] Arroyo MTK, Marticorena C, Miranda P, Matthei O, Landero A, Squeo F. Contributions to the high elevation flora of the Chilean Patagonia: A checklist of species on mountains on an East-West transect in the Sierra de los Baguales, Latitude 50°S. Gayana Botanica. 1989;**46**:121-151

[24] Arroyo, MTK, Marticorena C, Muñoz M. A checklist of the native annual flora of continental Chile. Gayana Botanica. 1990;**47**:119-135

[25] Parodi LR. Revisión de las especies austroamericanas del género *Alopecurus*. Revista de la Facultad de Agronomia y Veterinaria. 1931;**7**:345-369

[26] Peñailillo P. *Anatherostipa*, un nuevo género de *Poaceae* (Stipeae). Gayana Botanica. 1996;**53**:277-284

[27] Peñailillo P. Los géneros nativos de la tribu Stipeae (*Poaceae*, Pooideae) en Chile. Theoria. 1996;**14**(1):125-140

[28] Peñailillo P. El género *Jarava* Ruiz et Pav. (Stipeae-Poeae): Delimitación y nuevas combinaciones. Gayana Botanica. 2002;**59**:27-34

[29] Matthei O. Estudio crítico de las gramíneas del género *Stipa* en Chile. Gayana Botanica. 1965;**13**:1-137

[30] Rúgolo ZE. Revalidación del género *Bromidium* Nees & Meyen emend. Pilger (Gramineae). Darwiniana. 1982;**24**:187-216

[31] Matthei O. El género *Bromus* L. (*Poaceae*) en Chile. Gayana Botanica. 1986;**43**:47-110

[32] Parodi LR. Gramíneas austroamericanas nuevas o críticas. Notas del Museo De La Plata, Botanica. 1938;**3**(17):15-33

[33] Essi L, Longhi-Wagner HM, Souza-Chies TT. New combinations within the *Briza* complex (*Poaceae*: Pooideae: Poeae). Novon. 2011;**21**:326-330

[34] Gay C. Atlas de la Historia Física y Política de Chile. Imprenta E. Thunot y Ca. Paris:1854. 99 p.

[35] Parodi LR. Sinopsis de las gramíneas chilenas del género *Chusquea*. Revista Universitaria. 1945;**30**:61-71

[36] Blanco P, Puig-Samper MA. Plantas de R.A. Philippi (1808-1904) en el herbario de la Comisión Científica al Pacífico (1862-1866) del Real Jardín Botánico de Madrid. Anales del Jardin Botanica de Madrid. 1995;**53**:55-59

[37] Judziewicz EJ, Clark LG, Londoño X, Stern MJ. American Bamboos. Washington DC., USA: Smithsonian Institution; 1999. p. 392

[38] Arriaga MO, Gómez Cadret RA, Montero MA. Catálogo de los tipos depositados en el herbario del Museo Argentino de Ciencias Naturales "Bernardino Rivadavia" (BA). II. Monocotiledóneas: *Poaceae*. Revista del Museo Argentino de Ciencias Naturales. 2001;**3**:135-146

[39] McClure FA. Genera of Bamboos native to the New World (*Gramineae*: *Bambusoideae*). Smithsonian Contributions to Botany. 1973;**9**:1-148

[40] Skottsberg C. Derivation of the flora and fauna of Juan Fernández and Easter Island. The Natural History of Juan Fernández and Easter Island. 1956;**1**:193-439

[41] Matthei O. Las especies del género *Chusquea* Kunth (*Poaceae*: Bambusoideae) que crecen en la X Región, Chile. Gayana Botanica. 1997;**54**:199-200

[42] Clark LG. *Chusquea* Kunth. In: Judziewicz EJ, Soreng RJ, Davidse G, Filgueiras TS, Zuloaga FO, editors. Catalogue of New World Grasses (*Poaceae*): I. Subfamilies *Anomochloideae, Bambusoideae, Ehrhartoideae,* and *Pharoideae*. Washington DC: Smithsonian Institution; 2000. pp. 36-52

[43] Luebert F, Pliscoff S, editors. Sinopsis Bioclimática y Vegetacional de Chile. Santiago: Editorial Universitaria; 2006. p. 262

[44] Pacheco N. *Chusquea* (*Poaceae*) en el Parque Nacional Puyehue, Región de Los Lagos, Chile [Internet]. 2013. Available from: http://www.clorischile.cl [Accessed: 28-02-2017]

[45] Philippi RA. Descripción de las plantas nuevas halladas en la Cordillera Pelada i de algunas otras de la provincia de Valdivia. Anales de la Universidad de Chile. 1865;**27**:324-333

[46] Caro JS, Sánchez E. Novedades en *Cynodon* (Gramineae) de América. Darwiniana. 1972;**17**:510-526

[47] Peterson PM. *Cynodon* rich. In: Peterson PM, Soreng RJ, Davidse G, Filgueiras TS, Zuloaga FO, Judziewicz EJ, editors. Catalogue of New World Grasses, *Poaceae*: II. Subfamily Chloridoideae. Washington DC: Smithsonian Institution; 2001. pp. 59-63

[48] Baeza CM. Los géneros *Danthonia* DC. y *Rytidosperma* Steud. (*Poaceae*) en América: Una revisión. Sendtnera. 1996;**3**:11-93

[49] Parodi LR. Las gramíneas sudamericanas del género *Deschampsia*. Darwiniana. 1949;**8**: 415-475

[50] Chiapella J, Zuloaga FO. A revision of *Deschampsia*, *Avenella*, and *Vahlodea* (*Poaceae*, Poeae, Airinae) in South America. Annals in the Missouri Botanical Garden. 2010;**97**:141-162

[51] Escobar I, Ruiz R, Finot VL, Negritto MA, Baeza CM. Revisión del género *Eragrostis* Wolf en Chile, basado en análisis estadísticos multivariados. Gayana Botanica. 2011;**68**:47-81

[52] Ospina JC, Aliscioni SS, Denham SS. A revision of *Festuca* (Loliinae, Pooideae, *Poaceae*) in Chile. Phytotaxa. 2015;**223**(1):1-66

[53] Matthei O. El género *Festuca* (*Poaceae*) en Chile. Gayana Botanica. 1982;**37**:1-69

[54] Parodi LR. Revisión de las gramíneas sudamericanas del género *Hierochloe*. Revista del Museo De La Plata. Botany. 1941;**3**(14):15-33

[55] De Paula, ME. Novedades en el género *Hierochloe* (Gramineae). Boletín de la Sociedad Argentina de Botánica. 1974;**15**:393-402

[56] De Paula ME. Las especies del género *Hierochloe* (Gramineae) de Argentina y Chile. Darwiniana. 1975;**19**:422-457

[57] Schouten Y, Veldkamp JF. A revision of *Anthoxanthum* including *Hierochloe* (Gramineae) in Malesia and Thailand. Blumea. 1985;**30**:319-351

[58] Connor HE. Flowers and floral biology of the holy grasses (*Hierochloe* and *Anthoxanthum*: Aveneae, Gramineae). Flora. 2012;**207**:323-333

[59] Nicora EG. Gramineae. In: Correa MN, editor. Flora Patagónica 3. Buenos Aires: Colección Científica del INTA; 1978. pp. 1-563

[60] Seberg O, Petersen G. Hordeeae Martinov. In: Zuloaga FO, Rúgolo ZE, Anton AM, editors. Flora Argentina 3(II). Córdoba, Argentina: Gráficamente Ediciones; 2012. pp. 51-83

[61] Acevedo M. Andropogoneae de la flora chilena (Gramineae). Boletín de la Sociedad Argentina de Botánica. 1968;**12**:350-372

[62] Peñailillo P. El género *Jarava* Ruiz et Pav. (*Stipeae-Poaceae*): Delimitación y nuevas combinaciones. Gayana Botanica. 2002;**55**:85-88

[63] Matthei O. El género *Megalachne* Steud. (Gramineae). Boletin de la Sociedad de Biología de Concepción. 1974;**68**:165-172

[64] Muñoz-Schick M. Revisión de las especies del género *Melica* L. (Gramineae) en Chile. Boletin del Museo Nacional de Historia Natural, Santiago de Chile. 1985;**40**:41-89

[65] Cialdella AM, Muñoz-Schick M, Morrone O. Sinopsis de las especies austro-americanas del género *Nassella* (*Poaceae*, Pooideae, Stipeae). Darwiniana n.s. 2013;**1**:76-161

[66] Muñoz-Schick M. Revisión del género *Nassella* (Trin.) E. Desv. (Gramineae) en Chile. Gayana Botanica. 1990;**47**:9-35

[67] Barkworth M, Torres MA. Distribution and diagnostic characters of *Nassella* (*Poaceae*: *Stipeae*). Taxon. 2001;**50**:439-468

[68] Catanzaro MP, Bonasora MG, Speranza PR, Medina-Nicolás M, Valls JFM, Rúa G. *Paspalum chilense* (*Poaceae*, Paspaleae): A new species from southern South America. Phytotaxa. 2015;**197**:245-256

[69] Parodi LR. El género *Phalaris* en Chile. Revista Argentina de Agronomia. 1939;**6**(2):76-84

[70] Finot VL. Taxonomía del género *Phalaris* L. (*Poaceae*: Pooideae: Phalaridinae) en Chile. Gayana Botanica. 2014;**71**:246-258

[71] Baldini RM. Revision of genus *Phalaris* L. (Gramineae). Webbia. 1995;**49**:265-329

[72] Desvaux EE. Gramíneas. In: Gay C, editor. Historia Física y Política de Chile, Botánica 6. Imprenta E. Thunot y Ca. Paris:1854.

[73] Parodi LR. Las especies del género *Phalaris* de la flora chilena. Revista Chilena de Historia Natural. 1943;**45**:130-136

[74] Cialdella AM, Arriaga M. Revisión de las especies sudamericanas del género *Piptochaetium* (*Poaceae*, Pooideae, Stipeae). Darwiniana. 1998;**36**:107-157

[75] Soreng RJ, Peterson PM. New records of *Poa* (*Poaceae*) and *Poa pfisteri*: A new species endemic to Chile. Journal of the Botanical Research Institute of Texas. 2008;**2**:847-859

[76] Giussani LM, Anton AM, Negritto MA, Romanutti AA, Soreng RJ. *Poa* L. In: Zuloaga FO, Rúgolo ZE, Anton AM, editors. Flora Argentina 3(II). *Poaceae*: Pooideae. Córdoba, Argentina: Gráficamente Ediciones; 2012. pp. 284-339

[77] Philippi RA. Observaciones sobre la flora de Juan Fernández. Anales de la Universidad de Chile. 1856;**13**:157-169

[78] Philippi RA. Remarques sur la flore de Île de Juan Fernández. Annales des Sciences Naturelles; Botanique, sér. 4. 1857;**7**:87-110

[79] Johow F. Estudios sobre la flora de las islas de Juan Fernández. Imprenta Cervantes. Santiago:1896. 289 p.

[80] Finot VL, Contreras L, Ulloa W, Marticorena A, Baeza CM, Ruiz E. El género *Polypogon* (*Poaceae*: Agrostidinae) en Chile. Journal of the Botanical Research Institute of Texas. 2013;7:169-194

[81] Finot VL, Peterson PM, Zuloaga FO, Soreng RJ, Matthei O. Revision of *Trisetum* (*Poaceae*: Pooideae: Aveninae) in South America. Annals of the Missouri Botanical Garden. 2005;**92**(4):533-568

Chapter 6

Beyond Turf and Lawn: Poaceae in This Age of Climate Change

Katherine Dunster

Additional information is available at the end of the chapter

http://dx.doi.org/10.5772/intechopen.69736

Abstract

Grassland ecosystems dominated by Poaceae are estimated to cover 40.5% of Earth's land base, and domestication of a few edible grass species into highly productive cereal grains aided the shift from nomadic food gathering to field cultivation and higher density habitation. In the Northern Hemisphere, grasses are used ornamentally and for pasture, fodder and forage with little thought that the grasses livestock grazed upon, or those we gaze upon have multi-functional uses elsewhere. In this age of climate change, the use of Poaceae solely for aesthetics and amenity depletes finite potable water supplies that are needed for human survival. Agricultural land is consumed for turf seed production, and land is removed from food production. Cereals and their growing regions, which we have depended upon as food for millennia, are unlikely to adapt to climate change, and this will result in food insecurity and famines. Despite traditional uses of Poaceae in various cultures, many unrealized needs for food, medicine and other material goods could be met elsewhere with knowledge transfer. Our modern relationship with grass as an ornamental or amenity must end. As designers, gardeners and urban dwellers, we must use Poaceae for its multi-functions, which will lead to resilience and survival.

Keywords: Poaceae, grass, lawn, aesthetics, food, medicine, technology, TEK, resilience, invasive, food security, climate change

1. Introduction

During the entire time that humans have existed on this planet, people have been central to landscape changes in most places—from small individual gestures to large collective efforts. Grassland ecosystems dominated by Poaceae are estimated to cover 40.5% of the land area of Earth, excluding Greenland and Antarctica [1]. Grasses though are found in Greenland [2], and more tenuously in Antarctica, where unintentionally introduced *Poa annua* L. has survived for more than a decade near a research station [3].

Grassland ecosystems and their soils gave rise to the coevolution of humans and edible grasses through hand selection and improvement of grains for food and beverages including rice, wheat, barley, corn, rye and oats [4, 5]. Dormant winter seasons in Northern Europe and the domestication of livestock led to the need and selection of superior grasses for hay, fodder and straw bedding [6, 7]. Other grasses proved useful for technology: fuel, shelter, clothing, rope, baskets and various domestic products.

The transformation from wild grain to domesticated cereal helped create high density places of human habitation, and humans require food, energy, water and wealth. Wealth gave rise to the use of grasses for aesthetics and recreation or leisure pursuits ranging from cosmetic lawn and ornamental gardening to turf for golf, lawn bowling, croquet and various other field sports. Until recently, human beings have generally had a free hand in shaping nature to fit our lifestyles. Humans have converted some 30% of the planet (about 3.8 billion hectares) to resource extraction, agriculture, urban and suburban uses. For example, turf grass and lawn coverage of the American landscape, while highly fragmented, was estimated at 164,000 km^2, which it is suggested, represents '*the single largest irrigated 'crop' in the US, occupying a total area three times larger than the surface of irrigated corn*' [8, p 3].

Conversion to lawn and turf in the US, based on urbanization rates, is increasing at an annual rate of 8000 km^2 [9], and these numbers do not take into account the agricultural areas required to produce grass seed to create lawns. The human desire for green, beautiful lawn requires considerable inputs that result in the use of scarce potable water, fertilizers that pollute potable water, and mowing produces greenhouse gases contributing to higher temperatures. Consequently, we now find ourselves facing planetary climate change that is not particularly human-friendly. In the urbanized world, the moment is long overdue where we need to think whether grass and lawn are an essential need, or just a need that follows long-established social norms, without questioning the impacts or necessity.

The planet is confronting dramatic changes in demographics, population growth and increasing frequency of severe climate impacts and natural disasters [10]. Planning and designing for resilience—environmental-social-economic—is the only way to help humans (mostly dwelling in cities) survive, adapt, grow and thrive as these changes affect demands on the built environment, infrastructure, transportation systems, and water and energy resources.

Basic to our actions are the concepts of human (social) adaptation, local and traditional knowledge, environmental values, place attachments and cultural landscapes. The entry point to action is understanding nature (including human nature), and what role it has in placing limits on, or even directing, our actions and efforts to adapt and become more resilient as humans. In the most developed countries, resilience means rejecting purely cosmetic grass use and embracing Poaceae for its traditional technologies within the urban aesthetic, exploring the domestication (or understanding the wild harvesting potential) of other species of Poaceae to adapt cultivation to meet changes in climate such as warmer temperatures, or the new normal of climate extremes.

As I describe in this paper, humans have barely touched the potential of Poaceae to meet their needs for food, medicine, and other cultural and material goods. Cereal qualities and traits were bred and selected to meet specific regional climate realities such as warm season, cool season, short day-length, flooding, higher yields, drought and pests, or disease tolerance. Unless the cultivar or species is locally adapted to changing climate, finding new species, landraces and cultivars that are suitable and will thrive in shifting weather patterns is urgent, to avoid food insecurity and famine at the worst [11, 12].

1.1. Poaceae and climate change

'Vulnerability hotspots', are described as regions likely to experience both a decline in adaptive capacity for wheat and maize, and a decline in available soil moisture [13, p. 195]. The regions with lowest adaptive capacity were identified as wheat production areas in western Russia, northern India, southeastern South America and southeastern Africa. Wheat regions most likely to be exposed to drought and lacking adaptive capacity were identified as southeastern USA, southeastern South America, the northeastern Mediterranean and parts of central Asia. For maize (corn), regions with the lowest adaptive capacity include the northeastern USA, southeastern South America, southeastern Africa and central to northern India; when drought is added to the maize model vulnerability hotspots identified include southeastern South America, parts of southern Africa and the northeastern Mediterranean. The future of these cereals as a reliable source of food in these regions, when greater yields and food security are imperative to feed growing populations planet-wide, is doubtful.

The effects of climate change on food production were modelled in West Africa, and a wide variety of adaptation options available were reviewed to ensure immediate and long-term food security, including selecting different cultivars and crop types, more inputs (water and fertilizer), water harvesting during rainy seasons for irrigation during dry periods, and zero-tillage [14, 15]. The adaptation option most likely to be successful under current climate conditions is to develop seeds with *'increased thermal resilience during grain formation'* [14, p. 304), which they note may sustain crop yields under present climate realities, but may not be the best adaptation option for future climate changes. This is the hallmark of a wicked problem [16], as characterized in **Table 1** [17].

1. You don't understand the problem until you have developed a solution; it's cumulative, cascading, synergistic and evolving as it is explored.
2. There is no stopping rule; no definitive problem = no definitive solution.
3. Solutions are not right or wrong, true-or-false, good-or-bad but rather are better/worse, good enough/not good enough.
4. Each wicked problem is essentially unique and novel.
5. Every solution to a wicked is a 'one-shot operation'; it can't be replicated with the same outcomes for any other wicked problem.
6. There is no given alternative solution; potential solutions may be crafted, but more may not have been thought of, and more will be through shared knowledge and understanding.

Table 1. Six characteristics of wicked problems.

1.2. Human adaptation and resilience to changing climate

Rather than dwell on whether an unsolvable wicked problem with potentially dire outcomes for West Africa has been described [14], it seems prudent to recognize that it may be wicked and that there will be parallel scenarios playing out elsewhere on this planet sharing the consequences of climate change. We know problems can't be solved without trying solutions, most often at great cost, including loss of life from famine. In the case of food security and climate change in vulnerable regions, good enough/not good enough strategies still leave too much uncertainty in both farm fields and dinner bowls.

Meanwhile, the privileged developed world needs to re-assess their relationship with grass simply as garden bling or amenity, and learn to associate grass with edible cereals, multi-functionality and survival. There is, for example, an urgent need to reduce and transition from turf seed production, the majority of which is produced in Oregon, to growing sufficient and locally specific climate-adapted cereals [18]. This would remove the burden of, for example, the US importing cereal products from climate-vulnerable countries, while reducing pressure on its own low climate-adaptive southeastern region, which will become an unreliable source of wheat in extreme drought episodes. Despite the US being a major wheat exporter, *Farming Monthly National* [19] reported the export of 63,000 metric tonnes of feed wheat, that is, wheat to feed livestock, from the UK to the US in 2016.

In forecasting shifting climate scenarios to 2070, it was found that Poaceae as a family is unlikely to adapt through climate niche change and migration to more amenable habitat [20]. This is particularly troublesome given that maize, rice and wheat are not only the major plants cropped globally, they currently account for 89% of all cereal production, and supplied 42% of all the calories consumed by humans in 2009 [21]. Instead, seed saving and banking of land-races and non-domesticated species, combined with assisted migration that mimics natural range expansion to safe sites, may become an active management strategy to protect biodiversity, and the food potential of species for the future [22]. It appears crucial to the search for more robust species from the wild to avoid the genetic bottlenecks that occurred through polyploidization events in the domestication of wheat [23]. Before the selection of a few species that led to complete domestication, experimentation with multiple grass species occurred in multiple places across the Fertile Crescent region for centuries [24]. Even then, at the end of the day, looking for new grasses with the potential to replace the staple grains of the past 20,000 plus years may be an exercise in futility (or another wicked problem), if the survival of all Poaceae in these times of climate change is threatened [20].

The challenge is further exacerbated by the conversion of global grassland ecosystems to other land uses, resulting in their degradation, and loss to urbanization [25]. Examples from dry grassland ecosystems on various continents provide a somewhat daunting perspective on the magnitude of effort required to repair landscapes needed for various ecosystem services, including food production [26]. And for additional insight regarding scale, over a period of 12 years more than 500,000 kg of seed from around 250 species was harvested to restore 90 km^2 of Minnesota tallgrass prairie, and that a typical year in which 1000 hectares were replanted '*required roughly 13,000 kg of seed, approximately 5% (640 kg) of which was hand collected*' [27, p. 3075].

Resilience then, in this age of climate change, means that if plants cannot adapt fast enough, humans have to step up and adapt their behaviours, personally and collectively. Every food security issue must be addressed globally, as is the United Nations in attempting to mitigate the present Sudan Famine crisis which has displaced 1.6 million people from East Africa [28]. All resilience and survival efforts must be multi-directional, multi-pronged, shared and collaborative, to simultaneously explore multiple solutions that might fit multiple scenarios, rather than locking in to one expensive alternative that may not eventually work in one region, but may show promise in another.

And so, in juxtaposing the use of grasses solely for ornamental or aesthetic values with famine and food insecurity, it is clear that the moral imperative must yield to seeking and dedicating land to growing high-yielding cereal grasses first and foremost for their food, medicinal and nutritional values. If by chance such grasses have additional values that might meet aesthetic, social and recreational needs, then all-the-better, as in the past, this was commonplace as documented for one area of the Mediterranean [29], or in the cottage gardens that evolved in Britain.

2. Historical perspective on Poaceae as food

> *Certain cereals and pulses (legumes) were domesticated in very ancient times. In about 8000 BC in the Fertile Crescent of the Near and Middle East (present-day Syria, Iran, Iraq, Turkey, Jordan, Israel), wheats, barley, lentil, pea, bitter vetch, chickpea, and possibly faba bean, were brought into cultivation by the Neolithic people. These crops spread from the point of origin. Archaeological evidence indicates that the wheats, and some of the legumes, had reached Greece by 6000 BC and evidence of their presence within that millennium has been found in the Danube Basin, the Nile valley, and the Indian subcontinent (Pakistan). Dispersal continued through Europe, the crops reaching Britain and Scandinavia in 4,000-2,000 BC. There was quite a hiatus in this dispersal until the sixteenth and following centuries when, following the exploration and colonization of various countries, wheat species were taken to North and South America, South Africa, Australia, and New Zealand.* [30, p. xxviii]

During the Paleolithic (Stone Age), which began around 2.5 million years BCE and lasted until the global advent of agriculture in various unglaciated places around 10,000 years ago, wild grass seed from many species was gathered from the landscape for food [5, 31]. It has been previously thought that '*seeds and beans were rarely eaten and never in large amounts on a daily basis*' in the Paleolithic [32, p. 75]. Recent archaeology at Paleolithic sites in Southern Italy has now found evidence that by 32,600 BCE, hunter-gatherers were gathering substantial quantities of wild grains, primarily temperate cool climate *Avena* (oat) species, and had devised stone pestle-grinder tools for the conversion of the grain to flour [33].

Around 8000 BCE, the process of selecting best-performing grains led to the domestication of some cereal crops that were cultivated in the Middle East, and then distributed westwards into Africa, and eastwards to South Asia. As glacial ice retreated, domesticated grains moved north into Europe to replace those wild gathered from around 5500 to 5000 BCE [32, 34]. About the same time that wheats (einkorn [*Triticum monococcum* L.], emmer [*T. dicoccum* Schrank ex Schübl.] and barley [*Hordeum vulgare* L.]) were domesticated, domestication of two other Poaceae, rice (*Oryza sativa* L.) in China and maize (*Zea mays* L.) in Mesoamerica, was also occurring [24]. Although Poaceae is but fifth largest of the plant families, the top four food plants in the world are from Poaceae (in ranked order): sugar cane (*Saccharum officinarum* L.), maize, wheat (*T. aestivum* L.) and rice [35].

While Poaceae currently has around 12,000 species, the majority have never been domesticated, and remain wild [36]. Though undomesticated, the harvesting of wild grass seeds has occurred on all inhabited continents [37], and persists right up to the present day in many parts of the world, even though seed collection and processing can be challenging for various reasons such as widespread distribution patterns with low abundance, shattering, lodging or competition with more dominant species. People turn to gathering wild grass seeds for various reasons—as a basic survival strategy to counter famine, poverty or economic depression, to maintaining traditional agricultural practices, preserving the traditional recreational or cultural activities of gathering of local wild food and plant medicines [29, 38–40], or for ecosystem restoration and biodiversity conservation purposes [41–43].

Of those species domesticated, grasses can be loosely categorized as edible (e.g. cereal grains), medicinal, ornamental, pasture (e.g. fodder and forage), turf (lawn), technological (e.g. biofuel, building, paper, clothing, oils, perfumes and craft materials) and spiritual (e.g. ceremonial smudging and smoking, incense, and other cultural rituals). Some grasses fall into several categories, and some grasses may have a domestic use unique to a single place, while considered as useless elsewhere [44]. Other native grasses domesticated somewhere for food, medicine or technology are dismissed outright as problematic weeds or invasive species somewhere else, with little regard for their rich histories, and traditional uses in their places of origin.

3. Loss of plant utility following translocation of a species

'Plant names often reflect people's belief systems and oral histories' [45, p. 1.171]. For example, quackgrass, originated in Europe and Central Asia, and was introduced to eastern North America by settlers in the eighteenth century [46]. It has since spread and naturalized urban and rural landscapes, been declared a weed in many states and provinces, and is the subject of substantial investments in time, money [47] and research to develop control methods that range from chemical pesticides to organic approaches using cover crops to prevent the spread [48].

Turn to any North American print or web reference on the weedy 'problem' of quackgrass—*Elymus repens* L., (syn. *Agropyron repens* L., *Elytrigia repens* (L.) Desv. ex Nevski), and the various common English names reflect beliefs and opinions about this plant that are relatively unflattering: couchgrass, twitch, quick grass, quitch grass, quitch, dog grass, quackgrass, scutch grass and witchgrass. However, in Lukomir, Bosnia and Herzegovina, for example, it is known as *piriki* [49], which, though used medicinally, translates from Bosnian (and also Croatian), to English as 'wheatgrass' [49]. *Piriki* then is a verbal reminder that this plant has utility as a cereal, albeit a poverty grain, that might be gathered in times of trouble such as the Bosnian War (1992–1995), when cultivating traditional cereal crops was diminished. Along with *vodenica mlini*, locally developed hydro-powered mills in Lukomir are supplied with water from wooden flumes to turn grindstones. When agriculture is practiced, cereal grains such as wheat, oats (*Avena sativa* L.), rye (*Secale sereale* L.), barley and corn are grown for flour.

In other parts of Europe, quackgrass was considered an important survival food during the First World War when seeds and rhizomes were ground into flour as a substitute for wheat and rye [46], while in Australia the rhizomes are ground sometimes into survival bread flour [50]. Before the First World War, it is reported that the mucilage exuded from quackgrass roots was as effective as glue that the United States imported a quarter-million pounds from Europe annually [51]. While quackgrass is not indigenous to North America, it soon became naturalized and, for example, the plants were used by the Okanagan-Colville peoples as a type of pit cooking container [52], while the White Mountain Apache Tribe of Arizona used quackgrass seeds for food [53]. In Ladakh, a region in Jammu and Kashmir, the northern-most state in India, powdered quackgrass rhizomes are used to traditionally treat irritated bladders and promote urination [54].

4. From quackgrass to quackery and back

Through the twentieth century, as the science of chemistry and modern technology expanded, herbalists were denounced as 'quacks' and herbal medicines were replaced by the component chemical compounds synthesized in laboratories and industrial factories [55]. Traditional knowledge has eroded or disappeared, as reported, for example, in Bali [56], or was saved for times of great need to survive famine, war and natural disaster without modern props. The repression and loss of cultural and ecological memory instigated by the colonialization of Indigenous peoples in North America and elsewhere [57, 58] has also turned around and traditional ecological knowledge (TEK) informing herbal medicine is moving from the alternative fringe to a larger arena [59–61].

There is a strong tug between staying local and respecting traditional foods and medicines, and embracing the benefits of globalization through access to new, potentially more tasty or effective products. This dilemma is articulated both from the Eastern perspective, where practitioners of Asian botanical medicine suggest that those working in the West should use Western herbs [62], or whether Eastern traditional use rules apply in the West [63]. In the West (Europe), it is posited that Western herbal medicine in the United Kingdom refers to '*using plants largely native to Europe, within a philosophical tradition arising from European thought*', and to avoid both North American and Eastern plants and healing traditions [64, p. 165].

By the turn of the twenty-first century, this attitude began to turn around as, for example, 'super bugs' invading humans have developed antibiotic resistance, and plant pathogens have likewise developed resistance to pesticides [65–68]. Reductionism in modern science is proving to be less helpful in facing new challenges because it compartmentalizes complex topics and ignores TEK, when there is considerable urgency and essentiality in supporting '*the integration of methods and results from different approaches and levels of analysis*' [69, p. 466]. A strong case is made for the integration of evidence-based medicine with TEK in order to avoid reductionism, and understand the plant holistically and ecologically, instead of breaking it down into useable parts and extracts [70]. Preserving TEK has significant positive implications for local socio-ecological resilience, and adaptation to change [71].

4.1. Ethnobotany, traditional use and the search for resilient grasses

The search for plants that have been traditionally used for food, technology and particularly medicine to replace ineffective modern creations has triggered a considerable amount of ethnobotanical field research across the planet. A plethora of research has been published, very recently, and I cite just a few here to provide some scope on the richness of information available for study [72–84]. Make no mistake that some of the studies cited are motivated by economics and profiteering, as the planet is scoured for 'new' plant materials that can become the 'next' food, nutraceutical, medicine or biotechnology product.

The drive to find better-yielding drought or flood-resistant cereals will tread heavily on the territory and cultures of Indigenous peoples. We must avoid the past mistakes of colonial resource exploitation, extraction and expropriation, or as Vandana Shiva says, 'biopiracy' [85]. Biopiracy is commonly seen as misappropriation or theft of plant genetic material—in part, or in whole. However, others speculate whether in the rush to find the 'next new', the motivation could be based on mutual aid and, with prior informed consent, results in knowledge transfer that helps bilateral economic development and conservation of species [86].

Prior informed consent requires the honouring of several UN international conventions. The *Convention on Biodiversity* [87], the *Universal Declaration on Cultural Diversity* [88], and most importantly, the *United Nations Declaration on the Rights of Indigenous Peoples* [89], which set forth global obligations to respect biodiversity and cultural diversity while protecting Indigenous rights to lands and intellectual property. If an outcome of plant exploration is the documentation of TEK, Indigenous peoples reserve their right to share, or not.

With the consent of residents, a study of home gardens in Iberia found TEK and agricultural knowledge to have blended with modern knowledge, resulting in greater social resilience to change [90]. This is attributed to personal changes acquired through learning new knowledge, practices and beliefs. An important point to note is that TEK is not static, but rather grows in response to new knowledge [91]. Rural to urban migration in China has had a negative impact on the environment as modern agriculture supplanted TEK, though through an experimental ecological education programme, the TEK of more ecological agricultural practices was successfully transferred back to the participating agricultural community [91]. Acceptance of TEK and Indigenous low-carbon living is advocated as being the key to climate change adaptation and resilience [92].

Climate change is an evidence-based wicked problem that may be incrementally and cumulatively solvable if we accept that we are all in it together, and admit that ingenuity may come from outside science. This requires solid doses of knowledge intersectionality, and humility, which requires recognizing and responding to ignorance in decision-making [93]. A final point to note is, '*that a specific unit of knowledge is lost or kept by a society is not as important as whether the society retains the ability to generate, transform, transmit, and apply knowledge*', which ultimately strengthens socio-ecological resilience [94, p. 646]. Good advice indeed as I continue this exploration of the changing role of Poaceae in this age of climate change.

5. The discounting of grass to single purposes and uses

In the 'Western' world particularly, and anywhere else that was exposed to European colonization, 'grass' is a generic term used to describe lawn, turf, sod and pasture. Where did these words originate, what do they really mean and why have they been essentially reduced to single purposes and uses? For the definitions and etymology of the 15 grass-related words shown in **Table 2**, I turned to the *Oxford English Dictionary* (OED) online [95].

Word	**Origins**	**First English use**	**Definition**
Turf	Old English (OE) turf	c725 CE	**1.** Slab pared from the surface of the soil with the grass and herbage growing on it; a sod of grass, with the roots and earth adhering. Also, in early quotations, a small portion of the sward *in situ*. **2.** A sod cut from the turf of an estate, and so on, as a token or symbol of possession.
Grass	OE græs, from Northern Europe	c725 CE	**1.** Herbage in general, the blades or leaves and stalks of which are eaten by horses, cattle, sheep, and so on. Also, in a narrower sense, restricted to the smaller non-cereal Gramineæ [sic], and plants resembling these in general appearance. **2.** The grassy earth, grass-covered ground; esp. ground covered with grass closely mown and rolled, forming a lawn in a public or private garden.
Hay	OE híeg, híg, hég, = Old Saxon houwi	c825 CE	Grass cut or mown, and dried for use as fodder; formerly (as still sometimes) including grass fit for mowing, or preserved for mowing.
Land	OE land	c900 CE	The solid portion of the earth's surface, as opposed to sea, water.
Fodder	OE foddor	c1225 CE	Food for cattle, horses or other animals. Now usually: hay, straw or other dried food used to feed animals, esp. in the winter. Also (U.S. regional): part or all of the corn plant used as animal food. Chaucer (c1390): 'Gras tyme is doon, my fodder is now forage'.
Meadow	OE mædewan	c1275 CE	A piece of land permanently covered with grass to be mown for use as hay; (gen.) a grassy field or other area of grassland, esp. one used for pasture. Also (regional): a tract of low well-watered ground, esp. near a river (cf. water meadow n.).
Pasture	Middle English (ME) pasturre, partly French pastour and Latin pastura	c1300 CE	A piece of grassy land used for or suitable for the grazing of animals, esp. cattle or sheep; pastureland. First as 'Oxpasture', a Yorkshire place-name.

Word	Origins	First English use	Definition
Forage	French fourrage	c1315 CE	Food for horses and cattle; fodder, provender; in early use, esp. dry winter food, as opposed to grass. Now chiefly provender for horses in an army.
Lawn	Old French launde	1340 CE	An open space between woods; a glade = laund, also meaning an open space among woods, a glade.
Herbage	ME before1500 French herbage, earlier 12th C. as 'erbage'	1390 CE	Herbs collectively; herbaceous growth or vegetation; usually applied to grass and other low-growing plants covering a large extent of ground, esp. as used for pasture.
Sod	Middle Dutch, sode, soode, Dutch zode	Before 1475 CE	A piece or slice of earth together with the grass growing on it, usually square or oblong in shape and of moderate thickness, cut out or pared off from the surface of grass land; a turf. Also const. of (grass, turf, etc.).
Swath I Swathe	OE swæþ, swaþu	c1475 CE	The space covered by a sweep of the mower's scythe; the width of grass or corn so cut.
Sward	OE sweard	1610 CE	A piece or slice of earth together with the grass growing on it, usually square or oblong in shape and of moderate thickness, cut out or pared off from the surface of grass land; a turf. To form a sward; to become covered with grassy turf.
Cereal	Latin Cereālis referring to Ceres, the goddess of agriculture	1818	Of or pertaining to corn or edible grain. A name given to those plants of the family Graminaceæ [sic] or grasses which are cultivated for their seed as human food; commonly comprised under the name corn or grain. (Sometimes extended to cultivated leguminous plants.)
Forb	Greek - φορβή fodder, forage (to feed)	1924	An herbaceous plant of a kind other than grass: applied chiefly to any broad-leaved herbs growing naturally on grassland.

Table 2. Words related to grass, extracted from the Oxford English Dictionary online [95].

Old English was spoken between the fifth and twelfth centuries in areas of what is now England and Southern Scotland, and it is from this period that several words related to grass made their appearance. Turf [96] and grass [97] are derived from Old English and make their way into our vocabulary around the same time. However, according to the OED they have never been combined into a single word viz., 'turf-grass' or 'turfgrass', which is commonly encountered in the industry. The second meaning of turf is related to its use in *seisin*, a c1300 feudal term for the conveyance of ownership, which required a witness to observe the physical and literal transfer of a small piece of ground from one person to another. From this meaning, it is easy to see how we talk about personal space in terms of 'my turf', or 'turf wars'.

While today, turf growers and managers would have you believe that turf is just grass or another word for lawn [98], it is clear from the definition that turf has always meant a slice cut from the surface of the land [99], including soil, grass and herbage (other low-growing plants). If only grass is growing in a slice of land, the words to use are sward [100] or sod [101]. Pasture [102] and grass are related to providing food and space for livestock to graze, while fodder [103] and forage [104] refer to swathes [105] of grass cut from meadows, dried and stored for winter feeding of livestock. A meadow [106] is a land permanently covered with grass (and other herbage [107] or forbs [108]) that is protected from grazing most of the year in order to harvest hay [109] to be used as winter fodder. Cereal only entered the lexicon in 1818, as the term for edible grains harvested from various grass species [110].

6. The meaning of lawn

While turf has retracted in meaning, the term lawn has expanded from its earliest connotation in the 1300s to describe an open space between treed woods that may have been used for pasturing livestock. By 1674 CE, the OED [98] indicates a change in definition to *'a stretch of untilled ground; an extent of grass-covered land'* and a hundred years later, lawn again grew to mean, *'a portion of a garden or pleasure-ground, covered with grass, which is kept closely mown'*.

In his monumental *Gardener's Dictionary* [111], Philip Miller, the chief gardener at the Chelsea Physic Garden from 1722 to 1770, expanded on the meaning of lawn, which he described as follows:

> *a great Plain in a Park, or a spacious Plain adjoining to a noble Seat. As to the Dimensions of it, it should be as large as the Ground will permit; but never less, if possible, than thirty or forty Acres. As to the Situation of a Lawn, it will be best in the Front of the House, and to lie open to the neighbouring Country and not pent up with Trees.*

This larger meaning, which Miller expands upon for several pages, certainly coincides with the period-defining works of the English landscape architects Capability Brown (c.1715–1783) and Humphry Repton (1752–1818), who were employed by wealthy landowners to modernize old Medieval gardens and agricultural land into what the designers called 'landscape parks', which would front many country mansions and stately homes throughout Britain, and be admired and then copied elsewhere.

Miller's suggestion that the lawn be located at the front of the house was essentially a visual cue to all who passed by to take notice that the landowner was so rich that they did not need the space for pasturing livestock. Landowners with lawns could afford to do nothing with it except look at it, sit on it and take walks on it, while their livestock was pastured somewhere else on the estate. In another display of landowner wealth, labourers were employed to maintain the large acreages of lawn by hand-scything the grass, with some assistance from pasturing sheep to maintain bucolic aesthetics. Fortuitously, the Industrial Revolution overlapped (c1760–1840), and the 1830 invention of the reel lawn mower pulled by horses, along with subsequent refinements throughout the nineteenth century, gave rise to a push mower that could be used by anyone to maintain a lawn.

Alongside the Industrial Revolution, the many mechanized inventions allowed an agricultural revolution to occur, shifting surplus labourers from farms into city factories, where they could be workers, rise to management or provide specialty trades. Living standards changed, including a rising middle class that had acquired enough wealth to own a city or town home with space for a garden. In the country, farm labourers and rural villagers typically used the area outside their front doors for growing flowers, fruit, herbs and vegetables—the 'cottage garden', which provisioned a household with many necessities. In cities and towns, the working class transferred their cottage garden to allotments. The middle class copied the aristocracy and grew verdant front lawns using improved cultivars of native grasses which required mowing to maintain appearances. Weekly mowing became a ritual. Further inventions created automatic watering devices and the push mower, various hand tools for pulling weeds and by the end of the 1800s, the green, weed-free lawn was ubiquitous

> *Shall it be a lawn of one kind of grass, or of several? Shall it be pure crested dogstail, dwarf and verdant, or shall it be the sheep's fescue of the downs, or shall it be a mixture of pons and fescues and clover? Shall it have yarrow in it, or shall it be severely grass, and grass alone? These are questions which are not to be answered hastily, particularly the yarrow question. They sow yarrow at Kew, but it is true that yarrow is not a grass. But, then, no more is clover. If one could make seven or eight lawns, it might be easier to decide such difficulties.* [112, p. 11]

From the mid-1800s, the front lawn as a measure of success shifted to North America and elsewhere, along with grass-related sports such as golf, lawn tennis and croquet. However, the traditional lawn grasses used in Britain, such as those mentioned in the above quote, were not adapted to the various climates on the North American continent. The US Department of Agriculture searched the globe and by 1897 had determined that Bermuda grass (*Cynodon dactylon L.*) from Africa, bluegrass (*Poa pratensis* L.) and ryegrass (*Lolium perenne* L.) from Europe, several native and non-native fescues (*Festuca* spp.), and creeping bent grass (*Agrostis stolonifera* L.) could be used singly or blended to maintain a robust lawn through the growing season [113]. Rural acreages were converted to the turf and lawn seed industry to meet the growing population's need for grass. By the end of the twentieth century, mown lawns were the dominant expression of urban and suburban landscapes in Europe and North America

> *Consider the many special delights a lawn affords: soft mattress for a creeping baby; worm hatchery for a robin; croquet or badminton court; baseball diamond; restful green perspectives leading the eye to a background of flower beds, shrubs, or hedge; green shadows – 'This lawn, a carpet all alive/With shadows flung from leaves' – as changing and as spellbinding as the waves of the sea, whether flecked with sunlight under trees of light foliage, like elm and locust, or deep, dark, solid shade, moving slowly as the tide, under maple and oak. This carpet!* [114, p. 159]

Lawns are a completely human creation, composed of mono- or poly-cultures of Poaceae, with the occasional inclusion of other forbs and weeds, and are artificially maintained solely for human use [115]. Admittedly, lawns can evoke the feelings in the above quote, but lawns also characterize the globalization of horticulture by ignoring the local—native species, environment and true human needs for resilience in this age of changing climate. As an amenity, lawn and turf grass landscapes consume limited resources such as money, time, water and energy, but even more troubling, they require upkeep that produces side effects such as pesticide run-off into potable water supplies and greenhouse gas emissions into the atmosphere from lawn mower exhaust. Lawns do nothing for urban food production and food security. Concern for the environment and the future of the planet requires recognition that traditional lawns are an unsustainable indulgence that requires a significant re-thinking of open spaces large and small, and then substantial praxis to create a new sustainable urban landscape for the twenty-first century.

The movement towards a new urban landscape began in the 1980s with the quest to find alternatives to grass lawns. This gave rise to many 'how-to' publications [116–120], along with growing interest in local, native approaches to gardening through publications on meadow, grassland and prairie garden creation [121–124]. All these authors advocate for using low-growing and spreading ornamental cultivars and native species to create more naturalistic, sustainable and resilient landscapes that are lower in maintenance, use less water and create habitat. However, the emphasis remained on the aesthetic and ornamental, still often borrowing non-native species from elsewhere in order to make design statements [125–127].

7. Beyond the beautiful: realising the full potential of Poaceae in human resilience

Climate change brings with it benefits and opportunities, including the ability to use plants previously considered not hardy to a region. In 2016, I began testing plants on our newly installed Roof Ecosystem Research Lab that was designed to serve the needs of the Urban Ecosystem programme at Kwantlen Polytechnic University in Langley, BC, Canada (49° 6'33.71"N; 122°38'47.08"W, USDA Climate Zone 8b). On a roof, which receives and absorbs more heat than on the ground, we can test plants at soil temperatures more conducive to lower latitude climates. In doing so, we can gain insight into the potential for new outdoor food crops such as sub-tropical and tropical chilli peppers (*Capsicum* spp.) that were previously unsuitable for outdoor growing because of a shorter season, or cooler and wetter prevailing weather patterns.

One of the plants we grew was lemongrass (*Cymbopogon citratus* (DC.) Stapf.), a tropical perennial native to South and Southeast Asia where it is used as both food [128] and medicine, including antioxidant, anti-fungal, anti-bacterial and anti-inflammatory properties [129]. Lemongrass and its traditional uses have spread throughout the sub-tropical and tropical regions of the world, in part because of its many outstanding medicinal qualities. In most parts of North America, it is hardy in USDA Climate Zones 10 and 11, where it thrives in summer heat. Lemongrass is not frost tolerant and is thus treated in our region as a novelty annual ornamental with spiky leaves, aromatic scent and culinary flavour enhancer anywhere prone to frosts. To achieve perenniality in temperate climates, winter cover is required to protect lemongrass from heavy frosts. While our plants did not survive the killing frosts of 2016–2017, we will be trialling lemongrass again in 2017, and covering with a low poly-hoop house on the roof for overwintering.

In the fall of 2016, a plant materials course taught by the author explored indigenous species from the West Coast of North America, including Poaceae, and their various cultural properties including food, medicine and technology uses. This nascent examination of the properties of Poaceae triggered a search of the university plant database to examine species planted on campus strictly as ornamentals, or labelled as weeds, that are used for food, medicine or technology in their places of origin. I have selected five species from our campus landscape to present as brief case studies, and highlight how little we know about the Poaceae. As designers, horticulturists and gardeners, we must move beyond the ornamental and aesthetic, to understand and embrace the full potential of all plant families to contribute to human resilience from human climate change and the need for food, medicine or different technologies.

7.1. Case study 1: millets, fountaingrass *(Pennisetum spp.)*

Throughout human history, various species of millet have been cultivated for their cereal grains, and a wide variety of millet species is used because *'they tolerate a range of soil types, and have short-weather growing seasons'* [130, pp. 150–151]. Yet, from a horticultural perspective in the Northern Hemisphere, *Pennisetum* spp. are valued solely for their ornamental qualities and the focus on breeding has *'led to improvements such as more intense purple foliage color, disease resistance, and apparent sterility'* [131, p. 525]. While there are five species and numerous cultivars of the genus *Pennisetum* on our university campus, knowledge beyond their use as an ornamental is poorly developed, and sterility might prevent viable seed collection and use.

Purple ornamental millet *(P. glaucum* 'Purple Majesty') is grown for its deep purple leaves that spill from robust stalks and prolific flowers [132]. The cultivar was an All-America Selection Gold Medal winner in 2003 [133]. *P. glaucum* is known in agriculture as pearl millet, and it has been used primarily as a summer grazing and hay crop in the US [134], and incidentally that, *'the cattail-like flower spikes that we left on the plants became covered in tan, feathery seed that small birds such as wrens, finches, and sparrows flocked to'* [133, p. 12]. Habitat attributes aside, pearl millet is highly useful to humans too. The seeds are used raw or cooked like rice, ground into flour, malted, or fermented into beverages such as beer [135, 136].

7.2. Case study 2: cogongrass *(Imperata cylindrica* [L.] Beauv.)

In temperate regions, several cultivars of *I. cylindrica* have been selected for garden use as ornamental plants, including the red-leaved 'Red Baron', also known as Japanese blood grass. As a native species, cogongrass is found in African and Asian grasslands, but has since spread through human dispersal to Micronesia, Australasia, Europe, southeast USA and Mexico. It now has over 100 common names, and is estimated to cover 2,000,000 km^2 (including natural grasslands) of the planet, so much so that it has been ranked in the top 10 worst weeds on the planet by the International Union for the Conservation of Nature [137]. Initially, cogongrass was introduced to new places for soil erosion control on agriculture and reforestation sites. Because it has few environmental limitations, the grass dispersed far beyond human need, or seemingly ability to control. Considerable effort has, and is, being invested into finding management treatments including pesticides and bio-controls that have potential to eradicate it from a variety of ecosystems, for a variety of reasons [138, 139].

Despite the negative aspects of this species, many other uses have been reported including processing the stems for roof thatch, rope and paper-making. As a food, the rhizomes are used to make beer in Malaysia, and in Australia and China the rhizomes are chewed to extract a sweet juice [50]. Throughout Asia, the rhizome has been used in traditional medicine to treat a vast array of ailments, is antibacterial and a diuretic [140]. Recent research indicates that components of cogongrass show promise in the treatment of colorectal cancer [141], and other research has identified root components such as alkaloids and flavonoids [142], and isolated the secondary metabolites [143].

The essential oils of cogongrass are used in Ayurvedic medicine to treat various illnesses [144], and phytotoxic components of the essential oils have also proven effective in the control of other weed species [145]. While cogongrass has great spiritual and medicinal significance to Vedic cultures on the Indian sub-continent, it is suggested that more of the phytotoxic properties should be tested as alternatives to synthetic chemical treatments that result in negative impacts to the environment [146]. Because this plant spreads aggressively by rapidly growing rhizomes, there is potential to derive social value through mechanical management that harvests the roots for processing, rather than using herbicides to kill it outright, and harm the environment.

7.3. Case study 3: maiden grass (*Miscanthus sinensis* Andress)

Miscanthus sinensis was originally introduced to North America from Asia in the 1890s as an ornamental plant, and quickly escaped cultivation and thrived because it produces viable seed, tolerates colder temperatures, has few pests and diseases, and has low demands for water and nutrients [147]. Over 150 cultivars of *M. sinensis* have been introduced [125], of which 40 are generally available to horticulture, and 10 occur on our campus. Traditional uses for *Miscanthus* include roof thatching material and fodder in Japan, where it is known as 'susuki' [148, 149, 150]. It is known as a wild food plant in Korea, where flowers and spikelets are consumed raw [151].

Range expansion scenarios of *Miscanthus* due to climate change indicate that the species will move northwards in North America, Eastern Europe and Scandinavia [152]. *Miscanthus* is considered a weed in Asia, and it was only late in the twentieth century when the genus was evaluated for biofuel potential in Asia [153], following earlier work in Europe [153]. The same qualities that endear this plant as an ornamental have proven favourable for biofuel production.

M. sacchariflorus arrived in the US around the same time as *M. sinensis*, while in Denmark, sometime around 1935 they naturally crossed and formed a triploid (sterile) hybrid known as *M. x giganteus* (Greef and Deuter ex Hodkinson and Renvoize), which can grow to 3+ metres in height [149]. *M. x giganteus* is used in Europe as a commercial energy crop, providing heat and electricity, and ethanol biofuel [154]. The biofuel potential of *Miscanthus* in North America shows much promise as a non-food replacement in the production of ethanol [147, 154], and experimental biofuel production is also occurring in Asia [152]. Marginal lands, often infested with weeds and invasives, could be better managed if used for the production of biofuels such as *Miscanthus*, which, of all Poaceae, is the most productive and possibly the least destructive to the environment [155].

7.4. Case study 4: orchardgrass (*Dactylis glomerata* L.)

Orchardgrass is a tall perennial bunch grass with stiff flat-sided flower/seed heads that resemble a cock's foot, which became its common name in Britain [156]. The plant has coarse foliage which can grow to form dense tufts or tussocks if unimpeded by mowing and grazing. As a cool season species, it is native throughout most of Europe, temperate Asia and northern Africa. It is long established in other temperate regions, and is well adapted to areas with higher rainfall. As such, orchardgrass has long been favoured as a fodder and pasture grass [157], and is a ubiquitous reminder of the agricultural history of a region, long after the land has been converted to other uses.

Besides feeding livestock, orchardgrass has found several human uses. In Poland, orchardgrass is known as 'kupkówka' and the sweet stem base and inner part of young shoots were eaten as a children's snack [158]. In Hungary, it is known as 'ebir', and also eaten as a snack [80]. In south-east Turkey, orchardgrass is called 'ayrik', and after infusing in water, the decoction is consumed to treat rheumatism and urinary inflammations [79].

Nitrate and pesticide contamination of soil and ground water is a serious environmental problem, and finding solutions is the focus of much research. In one study, orchardgrass was the most effective grass tested to remove toxic organic chemicals produced by the leaching of creosote into soil [159]. The *in situ* efficacy of various grasses in remediating soils contaminated with military explosive residues such as TNT found that orchardgrass was most effective at taking up TNT [160]. The versatility of orchardgrass has been demonstrated through successful recovery of nitrogen from the large amount of manure waste applied to pastures by dairies [161].

Along with several other cool-season grasses, orchardgrass was found effective at degrading atrazine in the soil column, and would be a suitable ground layer species for riparian buffer plantings in conventional agricultural areas that are heavily treated with herbicides and other chemicals [162]. Orchardgrass has proven to be a highly useful plant in bio-remediation initiatives to address past environmental mistakes, and this is possibly its highest and best use as we take responsibility for our misuses of ecological goods and services.

7.5. Case study 5: reed canary grass (*Phalaris arundinacea* L.)

Reed canary grass is one of five native *Phalaris* species in North America. It is circumboreal in distribution and native to both North America and Europe where it occurs naturally in wetlands and on the margins of aquatic habitat where there are wet, poorly drained soils [163]. On both continents, it is an important forage grass, which led to the importation of superior seed from Sweden to Canada in the early 1900s [164]. Other cultivars, both agricultural and ornamental, were imported from Asia, and both European and Asian ecotypes have escaped and hybridized with native North American populations. As a consequence, both the escaped cultivars and the hybridized ecotypes exhibit far more aggressive behaviour than the native species, resulting in some confusion over what is native or not, and the listing of reed canary grass as an invasive plant in many US states and Canadian provinces [165, 166].

On our university campus, *P. arundinacea* is the dominant plant species on wet meadows and the unbuilt floodplain areas of a creek that meanders through. Reed canary grass serves as a benchmark by marking the upper edges of flooding and more severe rainstorm effects, which are caused by climate change, and exacerbated by upstream urbanization. Left unmanaged reed canary forms impenetrable 200-cm tall thickets. When rough mown several times a growing season, it functions as a cut meadow, which allows us to carry out riparian ecosystem repair projects that make the floodplain more resilient to flooding. Rather than dwell on whether it is native or not, we hold the species in check by removing the opportunity for seeds to disperse and spread further. At the same time, the rhizomes hold soil in place and prevent erosion. Our management technique has been to leave the cut grass to decay and add organic matter where it drops, which, as you will see in the subsequent text, may be wasteful.

Finding cultural and economic uses for plants such as reed canary grass is perhaps the most ecologically sensitive approach to managing invasiveness. Ecotypes of reed canary grass that are translocated to suitable habitat in other regions behave far more aggressively than the native ecotype [167, 168], and reed canary runs rampant and out-competes other species in the native ecosystem [169].

In First Nations plant technology in British Columbia, reed canary grass is traditionally used in weaving, and the harvest of materials was sustainable, allowing the plants to be judiciously managed without depleting the resource [170]. By harvesting while the plants are still green and growing in late-spring, the reed canary grass can be managed through the removal of flowering parts. The process involves several steps, timed to the seasons:

> *The Upper Sto:lo of the Fraser River, the Lowe Stl'atl'imx and probably other Salish groups imbricated coiled cedar-root baskets with the stout, smooth stems of Reed Canary Grass. They gathered pliable, green stems in May and early June, around the time when wild roses bloom, cut them into even lengths and soaked them in boiling water, then dried them in the sun for several days to bleach them white. They split the dried stems, soaked them, and used them, …., to superimpose white patterns on the weave of split-root baskets.* [171, p. 119]

By contrast, from the world of new technology, the potential use of reed canary grass as a short fibre material found that is more sustainable than the typical trees used in the pulp and paper industry [172]. In Finell's research, a delayed-harvest technique holds the harvest as late as possible until the biomass is completely dry, which may be as late as the following spring after snow-melt, but before new growth begins. The reason for this delay is that *'for fibre production, the delayed harvest gives higher pulp yield, less variation, and stronger fibres'* [172, pp. 19–20]. In drier interior climates, this technique would work; however, in rainy coastal climates, biomass decay is more likely, requiring a fall harvest.

In other cultures, reed canary grass has long been recognized for its medicinal properties, most notably as a psychoactive drug from ancient Greece when Dioscorides reported that *P. arundinacea* was crushed and mixed with water or wine to treat bladder diseases [173]. The natural hallucinogenic alkaloids dimethyl-tryptamine and 5-methoxy-dimethyltryptamine (DMT) can be extracted from reed canary grass [174], and while toxic to livestock, DMT has medical benefits to humans [175].

It has been pointed out that the Internet has made it possible for anyone to find information on growing, harvesting and processing medicinal plants such as reed canary grass for use as a hallucinogen [176]. Humans have a long historical and cultural relationship with the use of hallucinogenic plants to alter the spirit, mind and body [173]. The harvesting of reed canary grass for any cultural or technological purpose would help manage and diminish its aggressive invasiveness in sensitive ecosystems. Natural area managers should look to intersectional collaborations between the scientific approach, TEK and high-tech to address invasive grasses such as reed canary grass, which currently are tagged as wicked problems.

8. Concluding remarks

In Indigenous cultures throughout northern North America (US and Canada), native grasses are gathered for medicine, clothing, domestic products, technology uses and fuel. Of the nearly 900 native grasses found in North America north of Mexico [163], only two native genera in the Poaceae have been domesticated for food, wild rice (*Zizania* spp.) and maize (Zea mays) was domesticated about 7000 years ago from a wild ancestor (*Zea mexicana* (Schrad.) Kuntze), commonly called teosinte [177]. Traditional knowledge of plants and their uses has always been transferred from generation to generation through everyday life activities [178, 179].

As the world grows in population and becomes increasingly urbanized, the connection to Poaceae as a source of food, medicine and other uses is diminished, as is TEK transfer. The need to find cereals more resilient to climate change is increasing in order to feed growing populations and avoid famine. Grasses have been moved around the planet intentionally (for cereal production, pasture, forage, turf and lawn) and over 400 introduced grass species have been recorded in the US and Canada [163]. Many are deemed to be invasive or weedy species, yet they have adapted to their new environment, and in some cases, thrive better than in their native habitat. As I have described in the case studies, many grasses used for ornamental or agricultural purposes, or are declared weeds, have great human utility in their native regions that is poorly understood, if at all in their new locations.

Recognizing the utility of all Poaceae—for food, medicine, fuel—as we know from TEK, traditional science, and the historical record must become a priority for those using grass in urban design, whether for sports turf, front lawns or ornamental plantings. And lastly, humans need to determine whether their survival is dependent on grasses devoted to sports turf, front lawns or ornamental plantings, or whether those spaces should be used, for example, to cultivate edible and medicinal grasses, while the residues are converted to biofuel.

Author details

Katherine Dunster

Address all correspondence to: unfoldinglandscapes@gmail.com

School of Horticulture, Urban Ecosystems Program, Kwantlen Polytechnic University, Surrey, BC, Canada

References

[1] Suttie JM, Reynolds SG, Batello C. Grasslands of the World. Rome, Italy: UN Food and Agriculture Organization; 2005

[2] Feilberg J. A Phytogeographical Study of South Greenland: Vascular Plants. Meddelelser om Grønland, Bioscience 15. Copenhagen: Kommissionen for Videnskabelige Undersøgelser i Grønland; 1984

[3] Frenot Y, Convey P, Lebouvier M, Chown SL, Whinam J, Selkirk PM, Skotnicki M, Bergstrom DM. Antarctic and subantarctic biological invasions: Sources, extents, impacts and implications. In: Rogan-Finnemore M, editor. Non-native Species in the Antarctic, Proceedings, pp. 53-96. Gateway Antarctica Special Publication Series Number 0801. Christchurch, NZ: University of Canterbury; 2008. Available from: http://citeseerx.ist.psu.edu/viewdoc/download?doi=10.1.1.457.3086&rep=rep1&type=pdf

[4] Nadel DP, Piperno DR, Holst I, Snir A, Weiss E. New evidence for the processing of wild cereal grains at Ohalo II, a 23 000-year-old campsite on the shore of the Sea of Galilee, Israel. Antiquity. 2012;**86**(34):990-1003

[5] Cunniff J, Wilkinson S, Charles M, Jones G, Rees M, Osborne CP. Functional traits differ between cereal crop progenitors and other wild grasses gathered in the Neolithic Fertile Crescent. PLoS One. 2014;**9**(1):e87586. DOI: 10.1371/journal.pone.0087586

[6] Davies MS, Hillman GC. Domestication of cereals. In: Chapman G.P, editor. Grass Evolution and Domestication. Cambridge: Cambridge University Press; 1992. pp. 199-224

[7] deWet JMJ. The three phases of cereal domestication. In: Chapman GP, editor. Grass Evolution and Domestication. Cambridge: Cambridge University Press; 1992. pp. 176-198

[8] Milesi C, Elvidge D, Dietz JB, Tuttle BT, Nemani RR, Running SW. A strategy for mapping and modeling the ecological effects of US lawns. In: Moeller M, Wentz E, editors. Proceedings, WG VIII/1 Joint Symposia URBAN – URS, March 14-16, 2005, International Society for Photogrammetry and Remote Sensing Archives – Vol. XXXVI-8/W27. 2005. Available from: http://www.isprs.org/proceedings/XXXVI/8-W27/milesi.pdf

[9] Gu C, Crane III J, Hornberger G, Carrico A. The effects of household management practices on the global warming potential of urban lawns. Journal of Environmental Management. 2015;**151**:233-242. DOI: 10.1016/j.jenvman.2015.01.008

[10] IPCC. Climate Change 2014: Synthesis Report. Contribution of Working Groups I, II and III to the Fifth Assessment Report of the Intergovernmental Panel on Climate Change [Core Writing Team, Pachauri RK, Meyer LA, editors.]. Geneva, Switzerland: IPCC; 2014

[11] Khan ZR, Midega CAO, Pittchar JO, Murage AW, Birkett MA, Bruce TJA, Pickett JA. Achieving food security for one million sub-Saharan African poor through push-pull innovation by 2020. Philosophical Transactions of the Royal Society B. 2014;**369**(1639):20120284. pp. 11. DOI: 10.1098/rstb.2012.0284

[12] Ahmed KF, Wang G, Yu M, Koo J, You L. Potential impact of climate change on cereal crop yield in West Africa. Climatic Change. 2015;**133**(2):321-334. DOI:10.1007/s10584-015-1462-7

[13] Fraser EDG, Simelton E, Termansen M, Gosling SN, South A. 'Vulnerability hotspots': Integrating socio-economic and hydrological models to identify where cereal production may decline in the future due to climate change induced drought. Agricultural and Forest Meteorology. 2013;**170**:195-205. DOI: 10.1016/j.agrformet.2012.04.008

[14] Guan K, Sultan B, Biasutti M, Baron C, Lobell DB. Assessing climate adaptation options and uncertainties for cereal systems in West Africa. Agricultural and Forest Meteorology. 2017;**232**:291-305. DOI: 10.1016/j.agrformet.2016.07.021

[15] Cattivelli L, Rizza F, Badeck F-W, Mazzucotelli E, Mastrangelo AM, Francia E, Marèa C, Tondelli A, Stanca AM. Drought tolerance improvement in crop plants: An integrated view from breeding to genomics. Field Crops Research. 2008;**105**(1-2):1-14. DOI: 10.1016/j.fcr.2007.07.004

[16] Rittel HWJ, Webber MM. Dilemmas in a general theory of planning. Policy Sciences. 1973;**4**:155-169. DOI: 10.1007/bf01405730

[17] Conklin J. Dialogue Mapping: Building Shared Understanding of Wicked Problems. Chichester, England: John Wiley & Sons; 2006

[18] Giombolina KJ, Chambers KJ, Bowersox JW, Henry PM. From turf to table: grass seed to edible grains in the Willamette Valley. Journal of Agriculture, Food Systems, and Community Development. 2011;**2**(1):141-161. DOI: 10.5304/jafscd.2011.021.008

[19] Farming Monthly National. Largest UK wheat export to the USA departs. May 25, 2016. Available from: http://www.farmingmonthly.co.uk/news/arable/10561-largest-uk-wheat-export-to-the-usa-departs/

[20] Cang FA, Wilson AA, Wiens JJ. Climate change is projected to outpace rates of niche change in grasses. Biology Letters. 2016;**12**:20160368. DOI: 10.1098/rsbl.2016.0368

[21] GRiSP (Global Rice Science Partnership). Rice Almanac. 4th ed. Los Baños (Philippines): International Rice Research Institute; 2013

[22] Vitt P, Havens K, Kramer AT, Sollenberger D, Yates E. Assisted migration of plants: Changes in latitudes, changes in attitudes. Biological Conservation. 2010;**143**(1):18-27. DOI: 10.1016/j.biocon.2009.08.015

[23] Charmet G. Wheat domestication: Lessons for the future. Comptes Rendus Biologies. 2011;**334**:212-220. DOI: 10.1016/j.crvi.2010.12.013

[24] Brown TA, Jones MK, Powell W, Allaby RG. The complex origins of domesticated crops in the Fertile Crescent. Trends in Ecology and Evolution. 2008;**24**(2):103-109. DOI: 10.1016/j.tree.2008.09.008

[25] Andrade BO, Koch C, Boldrini II, Vélez-Martin E, Hasenack H, Hermann J-M, Kollmann J, Pillard VD, Overbeck GE. Grassland degradation and restoration: A conceptual framework of stages and thresholds illustrated by southern Brazilian grasslands. Natureza & Conservação. 2015;**13**(2):95-104

[26] Broadhurst LM, Jones TA, Smith FS, North T, Guja L. Maximizing seed resources for restoration in an uncertain future. BioScience. 2016;**66**(1):73-79. DOI: 10.1093/biosci/biv155

[27] Gerla P, Cornett M, Ekstein J, Ahlering M. Talking big: Lessons learned from a 9000 hectare restoration in the northern tallgrass prairie. Sustainability. 2012;**4**(11):3066-3087. DOI: 10.3390/su4113066

[28] UN (United Nations News Centre). South Sudan Now World's Fastest Growing Refugee Crisis – UN Refugee Agency. 17 March, 2017. Available from: http://www.un.org/apps/news/story.asp?NewsID=56367#.WNiOEme1vIU

[29] Gras A, Garnatje T, Bonet MÀ, Carrió E, Mayans M, Parada M, Rigat M, Valles J. Beyond food and medicine, but necessary for life, too: Other folk plant uses in several territories of Catalonia and the Balearic Islands. Journal of Ethnobiology and Ethnomedicine. 2016;**12**(1):23. DOI: 10.1186/s13002-016-0097-8

[30] Vaughn JG, Geissler CA. The New Oxford Book of Food Plants. 2nd ed. Oxford: Oxford University Press; 2009

[31] Willcox G, Fornite S, Herveux L. Early Holocene cultivation before domestication in northern Syria. Vegetation History and Archaeobotany. 2008;**17**(3):313-325. DOI: 10.1007/s00334-007-0121-y

[32] Lindeberg S. Palaeolithic diet ("stone age" diet). Scandinavian Journal of Food & Nutrition. 2005;**49**(2):75-77. DOI: 10.1080/11026480510032043

[33] Lippi MM, Foggi B, Aranguren B, Ronchitelli A, Revedin A. Multistep food plant processing at Grotta Paglicci (Southern Italy) around 32,600 cal BP. Proceedings of the National Academy of Sciences. 2015;**112**(39):12075-12080. DOI: 10.1073/pnas.1505213112

[34] Delcourt HR. The impact of prehistoric agriculture and land occupation on natural vegetation. TREE. 1987;**2**(2):39-44. DOI: 10.1016/0169-5347(87)90097-8

[35] Bennett BC. Plants as food. In: Bennett B, editor. Economic Botany. Encyclopedia of Life Support Systems (EOLSS); Developed under the auspices of the UNESCO. Oxford, UK: Eolss Publisher; 2010. Available from: http://www.eolss.net/sample-chapters/c09/e6-118-07.pdf

[36] Christenhusz MJM, Byng JW. The number of known plant species in the world and its annual increase. Phytotaxa. 2016;**261**(3):201-217. DOI: 10.11646/phytotaxa.261.3.1

[37] Harlan JR. Origins and processes of evolution. In: Chapman GP, editor. Grass Evolution and Domestication. Cambridge: Cambridge University Press; 1992. pp. 159-175

[38] Azam FMS, Biswas A, Mannan A, Afsana NA, Jahan R, Rahmatullah M. Are famine food plants also ethnomedicinal plants? An ethnomedicinal appraisal of famine food plants of two districts of Bangladesh. Evidence-Based Complementary and Alternative Medicine. 2014;**2014**:741712. pp. 28. DOI: 10.1155/2014/741712

[39] Dal Cero M, Salle R, Weckerle CS. The use of the local flora in Switzerland: A comparison of past and recent medicinal plant knowledge. Journal of Ethnopharmacology. 2014;**151**(1):253-264. DOI: 10.1016/j.jep.2013.10.035

[40] Söukand R, Hrynevich Y, Vasilyeva I, Prakofjewa J, Vnukovich Y, Paciupa J, Hlushko A, et al. Multi-functionality of the few: Current and past uses of wild plants for food and healing in Liubań Region, Belarus. Journal of Ethnobiology and Ethnomedicine. 2017;**13**(1):10. DOI: 10.1186/s13002-017-0139-x

[41] Morgan JP, Collicut DR, Thompson JD. Restoring Canada's Native Prairie's: A Practical Manual. Winnipeg, Manitoba: Prairie Habitats; 1995. Available from: http://www.naturenorth.com/RCNP/RCNP.pdf

[42] Gramineae Services Ltd. Recovery Strategies for Industrial Development in Native Prairie for the Dry Mixedgrass Natural Subregion of Alberta. Edmonton: Alberta Environment and Sustainable Resource Development; 2013. Available from: http://www.foothillsrestorationforum.ca/recovery-strategies-for-dry-mixedgrass

[43] Barr S, Jonas JL, Paschke MW. Optimizing seed mixture diversity and seeding rates for grassland restoration. Restoration Ecology. 2017;**25**(3):396-404. DOI: 10.1111/rec.12445

[44] Austin DF. Fox-Tail Millets (Setaria: Poaceae): Abandoned food in two hemispheres. Economic Botany. 2006;**60**(2):143-158

[45] Turner NJ. Ancient Pathways, Ancestral Knowledge: Ethnobotany and Ecological Wisdom of Indigenous Peoples of Northwestern North America. Vol. 2. Montréal and Kingston: McGill-Queen's University Press; 2014

[46] Kephart LW. Quack Grass. Farmers' Bulletin 1307, revised. Washington, D.C.: U.S. Department of Agriculture; 1931

[47] Werner PA, Rioux R. The biology of Canadian weeds. 24. *Agropyron repens*(L.) Beauv. Canadian Journal of Plant Science. 1977;**57**:905-919

[48] Bond W, Davies G, Turner RJ. The Biology and Non-Chemical Control of Common Couch (*Elytrigia repens*(L.) Nevski). 2007. Available from: http://www.gardenorganic.org.uk/sites/www.gardenorganic.org.uk/files/organic-weeds/elytrigia-repens.pdf

[49] Ferrier J, Saciragic L, Trakić S, Chen ECH, Gendron RL, Cuerrier A, Balick MJ, Red S, Alikadi E, Arnason JT. An ethnobotany of the Lukomir Highlanders of Bosnia & Herzegovina. Journal of Ethnobiology and Ethnomedicine. 2015;**11**(81). p. 81. DOI: 10.1186/s13002-015-0068-5

[50] Lim TK. Edible Medicinal and Non-Medicinal Plants: Vol. 12. Modified Stems, Roots, Bulbs. Heidelberg: Springer; 2016. DOI: 10.1007/978-3-319-26062-4

[51] AKEPIC—Alaska Exotic Plant Information Clearinghouse. Invasive Plants of Alaska. Anchorage, Alaska: Alaska Association of Conservation Districts; 2005

[52] Turner NJ, Bouchard R, Kennedy DID. Ethnobotany of the Okanagan-Colville Indians of British Columbia and Washington. Victoria: British Columbia Provincial Museum; 1980

[53] Reagan AB. Plants used by the White Mountain Apache Indians of Arizona. Wisconsin Archeologist. 1929;**8**:143-161. Available from: http://naeb.brit.org/uses/13047/

[54] Ballabh B, Chaurasia OP, Ahmed Z, Singh SB. Traditional medicinal plants of cold desert Ladakh—used against kidney and urinary disorders. Journal of Ethnopharmacology. 2008;**118**(2):331-339

[55] Leonti M, Verpoorte R. Traditional Mediterranean and European herbal medicines. Journal of Ethnopharmacology. 2006;**199**:161-167. DOI: 10.1016/j.jep.2017.01.052

[56] Sujarwo W, Arinasa IBK, Salomone F, Caneva G, Fattorini S. Cultural erosion of Balinese indigenous knowledge of food and nutraceutical plants. Economic Botany. 2014;**68**(4):426-437. DOI: 10.1007/s12231-014-9288-1

[57] Turner NJ, Ignace MB, Ignace R. Traditional ecological knowledge and wisdom of Aboriginal Peoples in British Columbia. Ecological Applications. 2000;**10**(5):1275-287. DOI: 10.2307/2641283

[58] Biróa É, Babai D, Bódis J, Molnár Z. Lack of knowledge or loss of knowledge? Traditional ecological knowledge of population dynamics of threatened plant species in East-Central Europe. Journal for Nature Conservation. 2014;**22**(4):318-325. DOI: 10.1016/j.jnc.2014.02.006

[59] Croom Jr, EM, Walker L. Botanicals in the pharmacy: New life for old remedies. Drug Topics. 1995;**139**(21):84. ProQuest document ID 205032093

[60] Bouldin AS, Smith MC, Garner DD, Szeinbach SL, Frate DA, Croom EM. Pharmacy and herbal medicine in the US. Social Science & Medicine. 1999;**49**(2):279-289. DOI: 10.1016/S0277-9536(99)00118-5

[61] Etkin N. Edible Medicines: An Ethnopharmacology of Food. Tucson: University of Arizona Press; 2006

[62] McIntyre M. The development of Chinese herbal medicine in a Western setting: A discussion paper. The Journal of Chinese Medicine. 2013;**102**(2013):58-64. Available from: http://go.galegroup.com.ezproxy.library.ubc.ca/ps/i.do?p=HRCA&sw=w&u=ubcolumbia&v=2.1&it=r&id=GALE%7CA339528956&sid=summon&asid=b8de9fb53aaf48f154a832b9957d3fe8

[63] Gilbert N. Herbal medicine rule book: Can Western guidelines govern Eastern herbal traditions? Nature. 2011;**480**(7378):S98–S99. DOI: 10.1038/480S98a

[64] Nissen N. Naturalness as an ethical stance: Idea(l)s and practices of care in western herbal medicine in the UK. Anthropology & Medicine. 2015;**22**(2):162-176. DOI: 10.1080/13648470.2015.1043789

[65] D'Costa VM, Griffiths E, Wright GD. Expanding the soil antibiotic resistome: Exploring environmental diversity. Current Opinion in Microbiology. 2007;**10**(5):481-489. DOI: 10.1016/j.mib.2007.08.009

[66] Abdallah EM. Plants: An alternative source for antimicrobials. Journal of Applied Pharmaceutical Science. 2011;**1**(6):16-20. Available from: http://japsonline.com/abstract.php?article_id=118

[67] Lucas JA, Hawkins NJ, Fraaije BA. The evolution of fungicide resistance. Chapter 2 in Advances in Applied Microbiology. 2015;**90**:29-92. DOI: 10.1016/bs.aambs.2014.09.001

[68] R4P (Reflection and Research on Resistance to Pesticides) Network. Trends and challenges in pesticide resistance detection. Trends in Plant Science. 2016;**21**(10):834-853. DOI: 10.1016/j.tplants.2016.06.006

[69] Mazzocchi F. Western science and traditional knowledge: Despite their variations, different forms of knowledge can learn from each other. EMBO Reports. 2006;**7**(5):463-466. DOI: 10.1038/sj.embor.7400693

[70] Evans S. Changing the knowledge base in Western herbal medicine. Social Science & Medicine. 2008;**67**(12):2098-2106. DOI: 10.1016/j.socscimed.2008.09.046

[71] Ruiz-Mallén I, Corbera E. Community-based conservation and traditional ecological knowledge: Implications for social-ecological resilience. Ecology and Society. 2013;**18**(4): 12. DOI: 10.5751/ES-05867-180412

[72] Lumpert M, Kreft S. Folk use of medicinal plants in Karst and Gorjanci, Slovenia. Journal of Ethnobiology and Ethnomedicine. 2017;**13**(1):16. DOI: 10.1186/s13002-017-0144-0

[73] Menendez-Baceta G, Pardo-de-Santayana M, Aceituno-Mata L, Tardío J, Reyes-García V. Trends in wild food plants uses in Gorbeialdea (Basque Country). Appetite. 2017;**112**:9-16. DOI: 10.1016/j.appet.2017.01.010

[74] Pieroni A, Soukand R, Quave CL, Hajdari A, Mustafa B. Traditional food uses of wild plants among the Gorani of South Kosovo. Appetite. 2017;**108**:83-92. DOI: 10.1016/j.appet.2016.09.024

[75] Tugume P, Kakudidi EK, Buyinza M, Namaalwa J, Kamatenesi M, Mucunguzi P, Kalema J. Ethnobotanical survey of medicinal plant species used by communities around Mabira Central Forest Reserve, Uganda. Journal of Ethnobiology and Ethnomedicine. 2016;**12**(5). DOI:0.1186/s13002-015-0077-4

[76] Yazdanshenas H, Shafeian E, Nasiri M, Mousavi SA. Indigenous knowledge on use values of Karvan district plants, Iran. Environment, Development and Sustainability. 2016;**18**(4):1217-1238. DOI: 10.1007/s10668-015-9698-y

[77] Abbbasi A, Khan MA, Shah MH, Shah MM, Pervez A, Ahmad M. Ethnobotanical appraisal and cultural values of medicinally important wild vegetables of Lesser Himalayas-Pakistan. Journal of Ethnobiology and Ethnomedicine. 2013;**9**(1):66. DOI: 10.1186/1746

[78] Kalle R, Sõukand R. Wild plants eaten in childhood: A retrospective of Estonia in the 1970s–1990s. Botanical Journal of the Linnaean Society. 2013;**172**(2):239-253. DOI: 10.1111/boj.12051

[79] Tetik F, Civelek S, Cakilcioglu U. Traditional uses of some medicinal plants in Malatya (Turkey). Journal of Ethnopharmacology. 2013;**146**(1):331-346. DOI: 10.1016/j.jep.2012.12.054

[80] Dénes A, Papp N, Babai D, Czúcz B, Molnár Z. Wild plants used for food by Hungarian ethnic groups living in the Carpathian Basin. Acta Societatis Botanicorum Poloniae. 2012;**81**(4):381-396. DOI: 10.5586/asbp.2012.040

[81] Cakilcioglu U, Khatun S, Turkoglu I, Hayta S. Ethnopharmacological survey of medicinal plants in Maden (Elazig-Turkey). Journal of Ethnopharmacology. 2011;**137**(1):469-486. DOI: 10.1016/j.jep.2011.05.046

[82] Mohagheghzadeh A, Faridia P, Shams-Ardakani M, Ghasemi Y. Medicinal smokes. Journal of Ethnopharmacology. 2006;**108**(2):161-184. DOI:10.1016/j.jep.2006.09.005

[83] Pieroni A, Price LL. Eating and Healing: Traditional Food as Medicine. New York, NY: Food Products Press; 2006

[84] Marles RJ. Aboriginal Plant Use in Canada's Northwest Boreal Forest. Vancouver: UBC Press; 2000

[85] Shiva V. Biopiracy: The Plunder of Nature and Knowledge. Boston, MA: South End Press; 1997

[86] Oldham P, Hall S, Forero O. Biological diversity in the patent system. PLoS One. 2013;**8**(11):e78737. DOI: 10.1371/journal.pone.0078737

[87] United Nations. Convention on Biological Diversity (with Annexes). No 30619. Rio de Janeiro, Brazil: United Nations; 1992

[88] UNESCO (United Nations Educational, Scientific and Cultural Organization). Universal Declaration on Cultural Diversity. Paris, France: UNESCO; 2002. http://unesdoc.unesco.org/images/0012/001271/127162e.pdf

[89] UN (United Nations General Assembly). United Nations Declaration on the Rights of Indigenous Peoples. New York, NY: United Nations; 2008. Available from: https://documents-dds-ny.un.org/doc/UNDOC/GEN/N06/512/07/PDF/N0651207.pdf?OpenElement

[90] Reyes-García V, Aceituno-Mata L, Calvet-Mir L, Garnatje T, Gómez-Baggethun E, Lastra JJ, Ontillera R, Parada M, Rigat M, Vallès J, Vila S, Pardo-de-Santayana M. Resilience of traditional knowledge systems: The case of agricultural knowledge in home gardens of the Iberian Peninsula. Global Environmental Change. 2014;**24**:223-231

[91] Tattoni C. Landscape changes, traditional ecological knowledge and future scenarios in the Alps: A holistic ecological approach. The Science of the Total Environment. 2017;**579**:27-36. DOI: 10.1016/j.scitotenv.2016.11.075

[92] Liu Y. Ecological education in rural China: Rediscovering traditional knowledge. Diaspora, Indigenous, and Minority Education. 2008;**2**(4):259-275. DOI: 10.1080/15595690802352846

[93] Raygorodetsky G. Why Traditional Knowledge Holds the Key to Climate Change. United Nations University; 13 December 2011. Available from: https://unu.edu/publications/articles/why-traditional-knowledge-holds-the-key-to-climate-change.html

[94] Rivera-Ferre MG, Ortega-Cerdà M. Recognising ignorance in decision-making: Strategies for a more sustainable agriculture. EMBO Report. 2011;**12**(5):393-397. DOI: 10.1038/embor.2011.55

[95] Gómez-Baggethun E, Reyes-García V. Reinterpreting change in traditional ecological knowledge. Human Ecology. 2013;**41**(4):643-647. DOI: 10.1007/s10745-013-9577-9

[96] Simpson JA, Weiner ESC. The Oxford English Dictionary. 2nd ed. [OED Online]. Oxford: The Clarendon Press; 1989. Available from: http://www.oed.com.ezproxy.library.ubc.ca/

[97] turf, n.1. [OED Online]. Oxford: Oxford University Press; February 2017. Web. 10 January 2017

[98] grass, n.1. [OED Online]. Oxford: Oxford University Press; February 2017. Web. 10 January 2017

[99] lawn, n.2. [OED Online]. Oxford: Oxford University Press; February 2017. Web. 10 January 2017

[100] land, n.1. [OED Online]. Oxford: Oxford University Press; February 2017. Web. 10 January 2017

[101] sward, n. [OED Online]. Oxford: Oxford University Press; February 2017. Web. 10 January 2017

[102] sod, n.1. [OED Online]. Oxford: Oxford University Press; February 2017. Web. 10 January 2017

[103] pasture, n. [OED Online]. Oxford: Oxford University Press; February 2017. Web. 28 February 2017

[104] fodder, n. [OED Online]. Oxford: Oxford University Press; February 2017. Web. 10 January 2017

[105] forage, n. [OED Online]. Oxford: Oxford University Press; February 2017. Web. 10 January 2017

[106] swath | swathe, n.1., n.2. [OED Online]. Oxford: Oxford University Press; February 2017. Web. 10 January 2017

[107] meadow, n. [OED Online]. Oxford: Oxford University Press; February 2017. Web. 20 February 2017

[108] herbage, n. [OED Online]. Oxford: Oxford University Press; February 2017. Web. 10 January 2017

[109] forb, n. [OED Online]. Oxford: Oxford University Press; February 2017. Web. 10 January 2017

[110] hay, n.1. [OED Online]. Oxford: Oxford University Press; February 2017. Web. 10 January 2017

[111] cereal, adj. and n. [OED Online]. Oxford: Oxford University Press; February 2017. Web. 10 January 2017

[112] Miller P. The Gardeners Dictionary: Containing the Methods of Cultivating and Improving the Kitchen, Fruit and Flower Garden, as also the Physick Garden, Wilderness, Conservatory, and Vineyard. Abridged from the folio edition. London. Printed for the Author. 1735

[113] Spectator Archive. English Lawns. 14 February 1908. p. 11. Available from: http://archive.spectator.co.uk/article/14th-march-1908/11/english-lawns

[114] Lamson-Scribner F. Lawns and lawn making. In: Yearbook of Agriculture for 1897. Washington, DC: Government Printing Office; 1898. Available from: https://naldc.nal.usda.gov/download/IND23343312/PDF

[115] White KS. Onward and Upward in the Garden. New York, NY: Farrar Straus Giroux; 1979

[116] Mellor DR. Picture Perfect: Mowing Techniques for Lawns, Landscapes, and Sports. Chelsea, Michigan: Ann Arbor Press; 2001

[117] Rubin C. How to Get Your Lawn and Garden off Drugs. Madeira Park: Harbour Publishing; 1989

[118] Rubin C. How to Get Your Lawn off Grass: A North American Guide to Turning Off the Water Tap and Going Native. Madeira Park: Harbour Publishing; 2002

[119] Stevie D. The Wild Lawn Handbook: Alternatives to the Traditional Front Lawn. New York, NY: Macmillan; 1995

[120] Primeau L. Front Yard Gardens: Growing More Than Grass. 2nd ed. Richmond Hill, ON: Firefly Books; 2010

[121] Hadden EJ. Beautiful No-Mow Yards: 50 Amazing Lawn Alternatives. Portland: Timber Press; 2012

[122] Greenlee J. The American Meadow Garden: Creating a Natural Alternative to the Traditional Lawn. Portland, Timber Press; 2009

[123] Lloyd C, Hunningher E. Meadows. Portland: Timber Press; 2004

[124] Steiner L Prairie-Style Gardens: Capturing the Essence of the American Prairie Wherever you Live. Portland: Timber Press; 2010

[125] Lewis P. Making a Wildflower Meadow. London: Frances Lincoln; 2015

[126] King M, Oudolf P. Gardening with Grasses. London: Frances Lincoln; 1998

[127] Quinn M, Macleod C. Grass Scapes: Gardening with Ornamental Grasses. North Vancouver: Whitecap Books; 2003

[128] Ardle J. Bamboos and Grasses. New York, NY: DK Publishing; 2007

[129] García-Casal MN, Peña-Rosas JP, Gómez-Malavé H. Sauces, spices, and condiments: definitions, potential benefits, consumption patterns, and global markets. Annals of the New York Academy of Sciences. 2016;**1379**(2016):3-16. DOI: 10.1111/nyas.13045

[130] Olorunnisola SK. Biological properties of lemongrass: An overview. International Food Research Journal. 2014;**21**(2):455-462. Available from: http://www.ifrj.upm.edu.my/volume-21-2014.html

[131] Barker G. Agricultural Revolution in Prehistory: Why did Foragers become Farmers? [Online]. Oxford: Oxford University Press; 2009. Available from: http://www.myilibrary.com?ID=223500

[132] Contreras RN, Owen J, Hanna W, Schwartz B. Evaluation of seven complex pennisetum hybrids for container and landscape performance in the Pacific Northwestern United States. HortTechnology. 2013;**23**(4):525-528. Available from: http://horttech.ashspublications.org/content/23/4/525

[133] Meyer MH. Ornamental grasses in the United States. In: Janick, J, editor. Horticultural Reviews. Vol. 39, Chapter 3. Hoboken, NJ: John Wiley and Sons; 2011. DOI: 10.1002/9781118100592.ch3

[134] Evans W. Ornamental millet. Organic Gardening. 2003;**50**(1):12. Academic Search Complete, EBSCOhost, search.ebscohost.com/login.aspx?direct=true&db=a9h&AN=9127335&site=ehost-live&scope=site

[135] Lee D, Hanna W, Buntin GD, Dozier W, Timper P, Wilson JP. Pearl Millet for Grain. Bulletin 1216. Statesboro, University of Georgia Cooperative Extension; 2012

[136] Tanaka T. Tanaka's Cyclopedia of Edible Plants of the World. Tokyo: Yugaku-sha (distributed by Keigaku Publishing Co.); 1976

[137] Obilana AO, Odhav B, Jideani VA. Functional and physical properties of instant beverage powder made from two different varieties of pearl millet. Journal of Food and Nutrition Research. 2014;**2**(5):250-257. DOI: 10.12691/jfnr-2-5-7

[138] GISD (Global Invasive Species Database). Species profile: Imperata cylindrica. 2017. Available from: http://www.iucngisd.org/gisd/speciesname/Imperata+cylindrica

[139] MacDonald GE. Cogongrass (*Imperata cylindrica)*– biology, ecology, and management. Critical Reviews in Plant Sciences. 2004;**23**:367-380. DOI: 10.1080/07352680490505114

[140] Overholt W, Hidayat P, Le Ru B, Takasu K, Goolsby JA, Racelis A, Cuda JP, et al. Potential biological control agents for management of cogongrass (Cyperales: Poaceae) in the Southeastern USA. Florida Entomologist. 2016;**99**(4):734-739. DOI: 10.1653/024.099.0425

[141] Duke JA, Ayensu ES. Medicinal Plants of China. Algonac, Michigan: Reference Publications, Inc.; 1984

[142] Kwok AHY, Wang Y, Ho WS. Cytotoxic and pro-oxidative effects of *Imperata cylindrica*aerial part ethyl acetate extract in colorectal cancer in vitro. Phytomedicine. 2016;**23**(5):558-565

[143] Jayalakshmi S, Patra A, Lal VK, Ghosh AK. Pharmacognostical standardization of roots of *Imperata cylindrica*Linn (Poaceae). Journal of Pharmaceutical Sciences and Research. 2010;**2**(8):472-476. Available from: http://www.jpsr.pharmainfo.in/Documents/Volumes/Vol2Issue8/jpsr%2002081005.pdf

[144] Chang I-F. Ecotypic variation of a medicinal plant *Imperata cylindrica*populations in Taiwan: Mass spectrometry-based proteomic evidence. Journal of Medicinal Plants Research. 2008;**2**(4):71-76. Available from: http://www.academicjournals.org/journal/JMPR/edition/April_2008

[145] Vijay SG, Ajay KM, Sannd R, Panda P, Rao MM. Evaluation of physicochemical parameters of *Imperata cylindrica*(Linn) Beauv. root used in Ayurvedic formulations. Research Journal of Pharmacy and Technology. 2012;**5**(10):1352-1355. Available from: http://www.indianjournals.com/ijor.aspx?target=ijor:rjpt&volume=5&issue=10&article=020

[146] Cerdeira AL, Cantrell CL, Dayan FE, Byrd JD, Duke SO. Tabanone, a new phytotoxic constituent of cogongrass (*Imperata cylindrica*). Weed Science. 2012;**60**(2):212-218. Available from: http://www.jstor.org.ezproxy.library.ubc.ca/stable/41497626

[147] Subramaniam S, Aravind A. Tradition to therapeutics: Sacrificial medicinal grasses *Desmostachya bipinnata*and *Imperata cylindrica*of India. Boletín Latinoamericano y del Caribe de Plantas Medicinales y Aromáticas. 2015;**14**(3):156-170. Available from: http://www.redalyc.org/toc.oa?id=856&numero=38535

[148] Quinn LD, Allen DJ, Stewart JR. Invasiveness potential of *Miscanthus sinensis*: Implications for bioenergy production in the United States. GCB Bioenergy. 2010;**2**(6):310-320. DOI: 10.1111/j.1757-1707.2010.01062.x

[149] Anzoua KG, Yamada T, Henry RJ. Miscanthus. In: Kole, C, editor. Wild Crop Relatives: Genomic and Breeding Resources, Industrial Crops. Chapter 9. Berlin and Heidelberg: Springer-Verlag; 2011

[150] Yamada T. Miscanthus. In: Cruz VMV, Dierig DA, editors. Industrial Crops. Handbook of Plant Breeding 9, Chapter 3; 2015. DOI:10.1007/978-1-4939-1447-0_3

[151] Song M-J, Kim H, Heldenbrand B, Choi KH, Lee B-Y. Traditional knowledge of wild edible plants on Jeju Island, Korea. Indian Journal of Traditional Knowledge. 2013;**12**(2):177-194. Available from: http://hdl.handle.net/123456789/16847

[152] Hager HA, Sinasac SE, Gedalof Z, Newman JA. Predicting potential global distributions of two Miscanthus Grasses: Implications for horticulture, biofuel production, and biological invasions. PLoS One. 2014;**9**(6):e100032. DOI: 10.1371/journal.pone.0100032

[153] Chung JH, Kim DS. Miscanthus as potential bioenergy crop in East Asia. Journal of Crop Science and Biotechnology. 2012;**15**(2):65-77. DOI: 10.1007/s12892-012-0023-0

[154] Lewandowski I, Clifton-Brown JC, Scurlock JMO, Huisman W. Miscanthus: European experience with a novel energy crop. Biomass and Bioenergy. 2000;**19**(4):209-227. DOI: 10.1016/S0961-9534(00)00032-5

[155] Heaton EA, Dohleman FG, Long SP. Meeting US biofuel goals with less land: The potential of Miscanthus. Global Change Biology. 2008;**14**(9):2000-2014. DOI: 10.1111/j.1365-2486.2008.01662.x

[156] Gopalakrishnan G, Christina NM, Snyder SW. A novel framework to classify marginal land for sustainable biomass feedstock production. Journal of Environmental Quality. 2011;**40**:1593-1600. DOI: 10.2134/jeq2010.0539

[157] Hughes GP. The role of cocksfoot (*Dactylis glomerata*) in grassland husbandry in Britain. Grass and Forage Science. 1962;**17**(3):225-228. DOI: 10.1111/j.1365-2494.1962.tb00301.x

[158] Darke R, Griffiths M, editors. The New RHS Dictionary Manual of Grasses. Portland, Oregon: Timber Press; 1994

[159] Łuczaj L, Szymański WM. Wild vascular plants gathered for consumption in the Polish countryside: A review. Journal of Ethnobiology and Ethnomedicine. 2007;**3**:17. DOI: 10.1186/1746-4269-3-17

[160] Rasmussen G, Olsen RA. Sorption and biological removal of creosote-contaminants from groundwater in soil/sand vegetated with orchard grass (*Dactylis glomerata*). Advances in Environmental Research. 2004;**8**(3-4):313-327. DOI: 10.1016/S1093-0191(02)00105-3

[161] Duringer JM, Craig AM, Smith DJ, Chaney RL. Uptake and transformation of soil [14C]-trinitrotoluene by cool-season grasses. Environmental Science and Technology. 2010;**44**(16):6325-6330. DOI:10.1021/es903671n

[162] Singer J, Moore K. Nitrogen removal by orchardgrass and smooth bromegrass and residual soil nitrate. Crop Science. 2003;**43**(4):1420-1426. DOI:10.2135/cropsci2003.1420

[163] Lin CH, Lerch RN, Garrett HE, George MF. Incorporating forage grasses in riparian buffers for bioremediation of atrazine, isoxaflutole and nitrate in Missouri. Agroforestry Systems. 2004;**63**(1):91-99. DOI:10.1023/B:AGFO.0000049437.70313.ef

[164] Barkworth ME, Anderton LK, Capels KM, Long S, Piep MB. Manual of Grasses for North America. Logan: Utah State University Press; 2007. Available from: https://muse.jhu.edu/

[165] Looman J. 111 Range and Forage Plants of the Canadian Prairies. Publication 1751. Ottawa: Agriculture Canada Research Branch; 1983

[166] GISD. 2017. Species Profile: *Phalaris arundinacea*. Available from: http://www.iucngisd.org/gisd/speciesname/Phalaris+arundinacea

[167] Invasives.org. 2017. Reed Canarygrass *Phalaris arundinacea*L. Available from: https://www.invasive.org/browse/subinfo.cfm?sub=6170

[168] Kercher S, Zedler J. Multiple disturbances accelerate invasion of reed canary grass (*Phalaris arundinacea*L.) in a mesocosm study. Oecologia. 2004;**138**(3):455-464. DOI: 10.1007/s00442-003-1453-7

[169] Spyreas G, Wilm BW, Plocher AE, Ketzner DM, Matthews JW, Ellis JL, Heske EJ. Biological consequences of invasion by reed canary grass (*Phalaris arundinacea*). Biological Invasions. 2010;**12**(5):1253-1267. DOI: 10.1007/s10530-009-9544-y

[170] Lavergne S, Molofsky J. Reed canary grass (*Phalaris arundinacea*) as a biological model in the study of plant invasions. Critical Reviews in Plant Sciences. 2004;**23**(5):415-429. DOI: 10.1080/07352680490505934

[171] Anderson MK. Tending the Wild Native American Knowledge and the Management of California's Natural Resources. Oakland: University of California Press; 2005

[172] Turner NJ. Plant Technology of First Peoples in British Columbia. Victoria: Royal BC Museum; 2007

[173] Finell M. The use of reed canary-grass (*Phalaris arundinacea*) as a short fibre raw material for the pulp and paper industry. [Doctoral dissertation]. Umeå: Swedish University of Agricultural Sciences; 2003. Available from: http://pub.epsilon.slu.se/378/1/Agraria_424_MF.pdf

[174] Ratsch C. The Encyclopedia of Psychoactive Plants: Ethnopharmacology and its Applications. Rochester: Park Street Press; 1998

[175] Barker RE, Hovin AW. Inheritance of indole alkaloids in reed canary grass (*Phalaris arundinacea*L), I: heritability estimates for alkaloid concentration. Crop Science. 1974;**14**(1): 50-53. DOI: 10.2135/cropsci1974.0011183X001400010015x

[176] Strassman RJ. Human psychopharmacology of N,N-dimethyl-tryptamine. Behavioural Brain Research. 1996;**73**(1-2):121-124. DOI: 10.1016/0166-4328(96)00081-2

[177] Halpern JH, Pope Jr, HG. Hallucinogens on the Internet: A vast new source of underground drug information. American Journal of Psychiatry. 2001;**158**(3):481-483

[178] Yang CJ, Kursel LE, Studer AJ, Bartlett ME, Whipple CJ, Doebley JF. A gene for genetic background in *Zea mays*: Fine-mapping enhancer of teosinte branched1.2 to a YABBY class transcription factor. Genetics. 2016;**204**(4):1573-1585. DOI: 10.1534/genetics.116.194928

[179] Kargioğlu M, Cenkci S, Serteser A, Evliyaoğlu N, Konuk M, Kök MŞ, Bağci Y. An ethnobotanical survey of inner-West Anatolia, Turkey. Human Ecology. 2008;**36**(5):763-777. DOI: 10.1007/s10745-008-9198-x

Section 4

Applied Circumstance of Grass Categories

Chapter 7

Lemon Grass (*Cymbopogon citratus*)

Miss Phool Shahzadi

Additional information is available at the end of the chapter

http://dx.doi.org/10.5772/intechopen.69518

Abstract

Lemon grass is, also called fever grass, a perennial plant with thin, long leaves that is indigenous to many Asian countries. Lemon grass contains citral, which is an essential oil, having medicinal and other useful significance. In the present work, essential oil 3,7-dimethyl-2,6-octadienal (citral) is hydrodistilled from lemon grass in the laboratory, 3,7-dimethyl-2,6-octadienal acetals (citral acetals) are synthesized from citral along with para-toluene sulfonic acid as a catalyst, which are used in perfumery, flavor, for fortifying lemon oil and has strong antimicrobial qualities. Infrared spectroscopy (IR) and gas chromatography (GC) were conducted for verification of chemical constitution present in essential oil and acetals of lemon grass. Nutritionally, lemon grass is a good source of vitamins A and C, folic acid, magnesium, zinc, copper, iron, potassium, calcium and manganese. Lemon grass oil (citral) is hydrodistilled and IR and GC are conducted to verify its constituents.

Keywords: lemon grass, essential oil, citral, ionons, vitamin E

1. Introduction

Lemon grass belongs to Cymbopogon, a genus of about 55 species of grasses, native to temperate and tropical regions. It is a lofty perennial grass. Common names of Cymbopogon include lemon grass, silky heads, citronella fever grass and barbed wire grass amongst many others. Essential oil called as citral or 3,7-dimethyl-2,6-octadienal is present in leaves and twigs of lemon grass which can be extracted easily by hydrodistillation. The essential oil of lemon grass has many important chemical constituents, which are helpful for many applications. It has cis and trans citral, myrcene, geranial, etc. Citral after distillation can be used for the synthesis of ionones, vitamin A, different types of citral acetals, these acetals has a wide range of applications in perfumery and helpful to reduce antibacterial activities.

1.1. Lemon grass oil

Lemon grass is a tropical herb of 3–6 feet length, leaves and twigs of this grass have essential oil, which has insect repellent activity. Leaves of this grass dried and stored for making tea, helps to cure many problems of stomach and anesthetic problems. Lemon grass leaves from local garden of PCSIR Laboratories were collected, dried under shade to deactivate starch, and cut into small pieces of 1–2 inches. Essential oil having lemon-like aroma was extracted by steam distillation, which can be used as scent and flavoring agents in medicine. It can help in fever reduction, helpful to improve digestion, reduce diarrhea, and stomachaches. As diluted oil, it is used to ease pain and arthritis, sterile stimulating, antispasmodic, and pain reliever. Lemon grass plant is shown in **Figure 1**.

1.2. Physical characteristics of citral

Name: 3,7-dimethyl-2,6-octadienal

Molar mass: 152.24 g/mol

Appearance: pale yellow liquid

Figure 1. Lemon grass plant.

http://dx.doi.org/10.5772/intechopen.69518

Odor: lemon like

Density: 0.893 g/cm^3

Boiling point: 229°C

Refractive index: 1.484–1.490

Lemon grass essential oil (citral, 3,7-dimethyl-2,6-octadienal) or lemonal belongs to monoterpenoids having a formula $C_{10}H_{16}O$. It is a diastereoisomer of E-isomer (geranial or citral A). The Z-isomer is known as neral or citral B.

Commercial citral, is as mixture of two isomers due to cis-trans isomerism at the C=C bond nearly the aldehyde group obtained from oils of plant sources. The geranial has strong lemon-like odor while neral has less. Citral can be used in cosmetic, medicine and food industries. It also has strong antimicrobial activities [1] and pheromonal effects in insects [2]. Citral is a basic intermediate for the synthesis of flavoring and fragrance components such as ionones, methyl ionones, and vitamins A [3] and E [4]. While the essential oil from lemon grass contains 70–75% of citral in addition to myrcene, geraniol, and nerol, is used in a cheap type of soap and cosmetics.

Structural Formula:

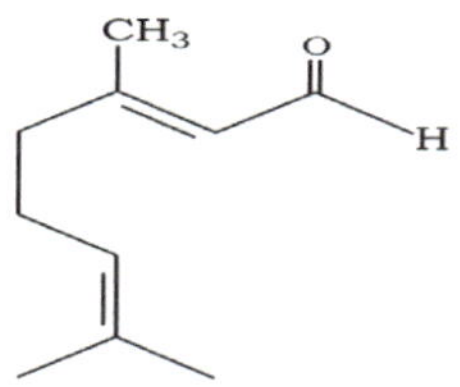

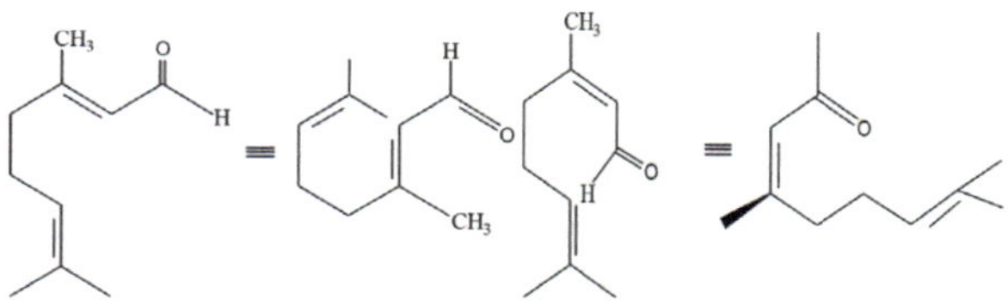

E- isomer (Geranial)
Trans (Citral- A)

Z-isomer (Neral)
Cis- (Citral - B)

1.3. The stereochemistry of geometrical isomers of citral

Ciral has two isomers, cis and trans. E-isomer (geranial) is trans and also known as citral A and Z-isomer (neral) is cis and known as citral B.

1.4. Sources of citral

There are two major sources of citral, which are explained in below.

1.4.1. Essential oils rich in citral

Percentage of oil (citral) varies according to plant species, about 90–98% oil is present in lemon myrtle, 70–80% in litsea cubeba, 65–85% in lemon grass, 30–35% in lemon verbena, 26% in ironbark lemon, 11% in lemon balm, 6–9% in lime, and about 2–5% in lemon and oranges [5].

1.4.2. Important plant sources of citral

Lemon myrtle contains chemotypes of two essential oils.

Lemon myrtle oil has typically 90–98% citral and oil yield 1–3% from fresh leaf. It is the highest natural source of citral.

The citronellal chemotype is uncommon and can be used as an insect repellant [6].

1.4.3. Litsea cubeba

Litsea cubeba fruits yields 3–5% essential oil. The 70–83% oil is obtained by oil's primary isolation. In China, oil production estimated between 500 and 1500 tonnes of oil per annum [7].

1.4.3.1. Lemon grass (Cymbopogon flexuosus)

Lemon grass (*Cymbopogon flexuosus*) [8], also called Cochin grass or Malabar grass, is native to India, Sri Lanka, Cambodia, Thailand, and Burma. *Cymbopogon citratus,* also known as serai in Malay, is assumed to have origin in Malaysia. While both can be used interchangeably, *C. citratus* is more used in cooking. In India, *C. citratus* is used both as a medical herb and perfumes.

1.4.4. Chemical synthesis of citral from pinene

Citral is also commercially produced from pinene [9]. The starting material is β-pinene (1) which is separated from turpentine oil by efficient fractional distillation of turpentine oil under vacuum. The β-pinene obtained in this way is passed through a heated tube with a short contact time at a temperature of about 600°C to furnish myrcene (2). The pyrolysis is thermal reaction and no catalyst is required.

Fortunately, the yield of myrcene under favorable condition can be as high as 90%. Myrcene is converted to the desired terpene mixture of cis and trans-alcohols, that is, geraniol and nerol by hydration of the double bond. This is not as easy as it would appear, to obtain a useful yield it is necessary to proceed via the hydrochloride, resulting a product, containing as much as 80% of linalyl chloride (3). The dry hydrochloric is passed through myrcene below −10°C in the presence of cuprous chloride. Linalyl chloride obtained in this way isomerizes into two isomers: geranyl chloride (4) and neryl chloride (5).

Either of the allylic chlorides can be converted to linayl or geranyl and neryl acetate by reaction with sodium acetate under suitable conditions. Thus if the sodium acetate is reacted in the presence of cuprous copper, the major product is linalyl acetate; whereas in the complete absence of copper, geranyl acetate (5) and neryl acetates (6) are predominate. The mixed terpene acetates undergo saponification to produce geraniol and nerol. These terpene alcohols are selectively oxidized to citral (7). As shown in **Scheme 1**.

http://dx.doi.org/10.5772/intechopen.69518

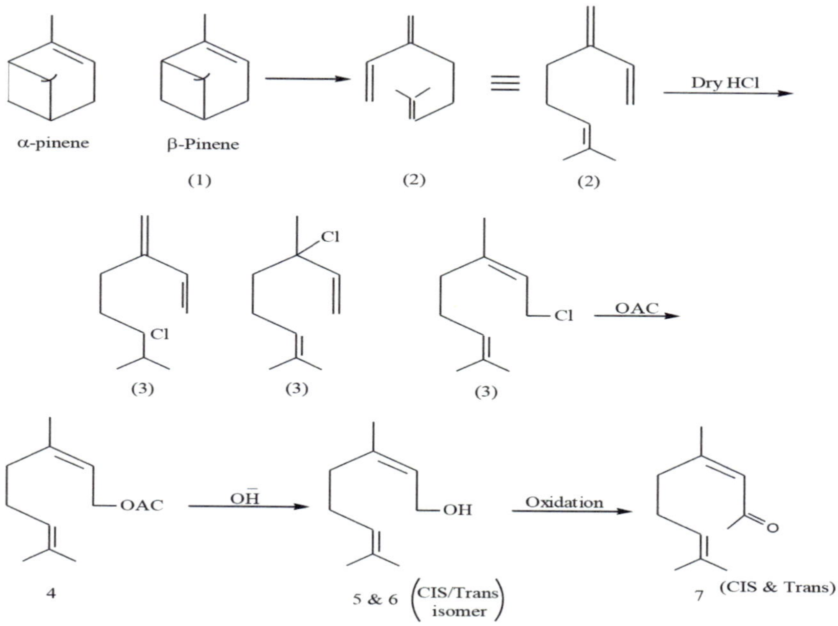

Scheme 1. Synthesis of citral from pinene.

1.5. Total synthesis of citral

BASF opened new continuous production of citral from 2-methyl- 4-hydoxy-but-1-ene in 2004 [10]. The annual production of citral is 40,000 tons per annum, which is shown in **Scheme 2**.

1.6. Synthetic applications of citral

1.6.1. Synthesis of ionones and methyl ionones from citral

The ionones belong to rose ketones having closely related series of chemical substances including damascones. A variety of essential oils contain these aroma compounds, the ionones, for example, β-ionone has rose aroma to some extent and it is used as a raw material for the production of retinol. Methyl-ionones are not found in the essential oils, but these are synthesized by aldol condensation of citral with methyl ethyl ketone (MEK).

(MEK).

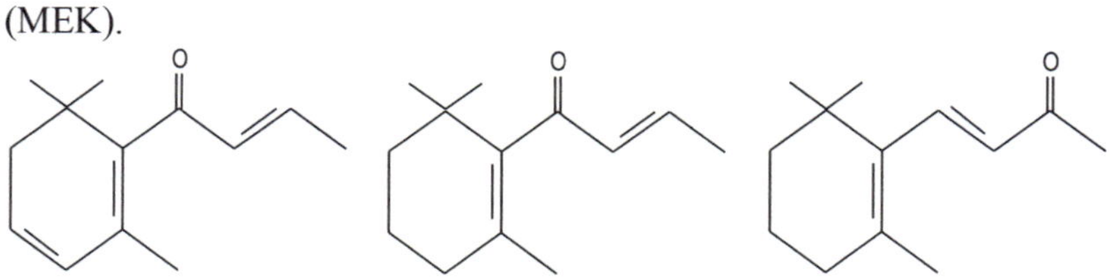

beta-Damascenone beta-Damascones beta-ionone

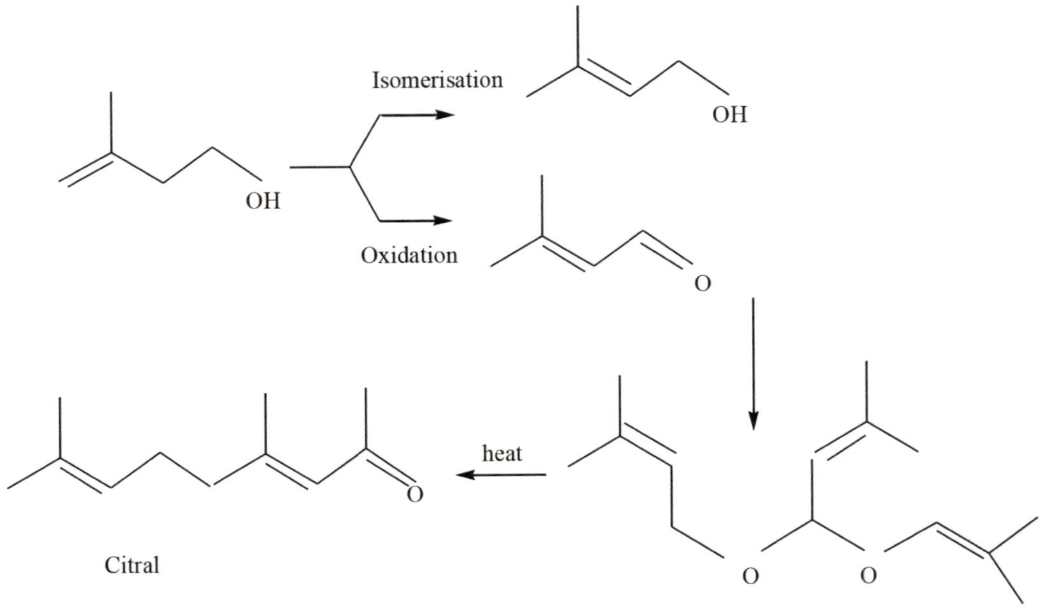

Scheme 2. Total citral synthesis.

1.6.2. Synthesis of ionones through pseudoionones

Ionone can be synthesized from citral though pseudoionone (PS). Citral and acetone by basic homogeneous or heterogeneous catalysis give PS, and serves as an example of aldol condensation. This is followed by a rearrangement reaction in acidic media to ionones. The nucleophilic addition of the carbanion (3) of acetone (1) to the carbonyl group on citral (4) is a base catalyzed. The aldol condensation product (5) eliminates water through the enolate ion (6) to the pseudoionone (7). The cyclisation reaction of PS to ionone proceeds by acid catalysis where the double bond in (7) opens to form the carbocation (8). A rearrangement reaction of the carbocation follows with ring closure to (10, 11). Finally, a hydrogen atom can be abstracted from (9) to form either (10) (extended conjugated system) or (11) as shown in **Scheme 3**.

Besides the pseudoionones, which are a mixture of two isomers as mentioned above, several side reactions can take place too, especially self-condensation of citral leading to dimerization and polymerization. Secondary reactions are potential problems. This is shown in **Scheme 4**.

1.6.3. Synthesis of vitamin A from ionone

Ionone is a key intermediate for the production of vitamin A. Ionone (C_{13} compound) on hydroformation gives C_{14} compound (2), which is treated with acetylene compound (a C_6 compound) to vitamin A having $C_{20,}$ by series of catalytic and isomerization reactions (see **Scheme 5**).

http://dx.doi.org/10.5772/intechopen.69518

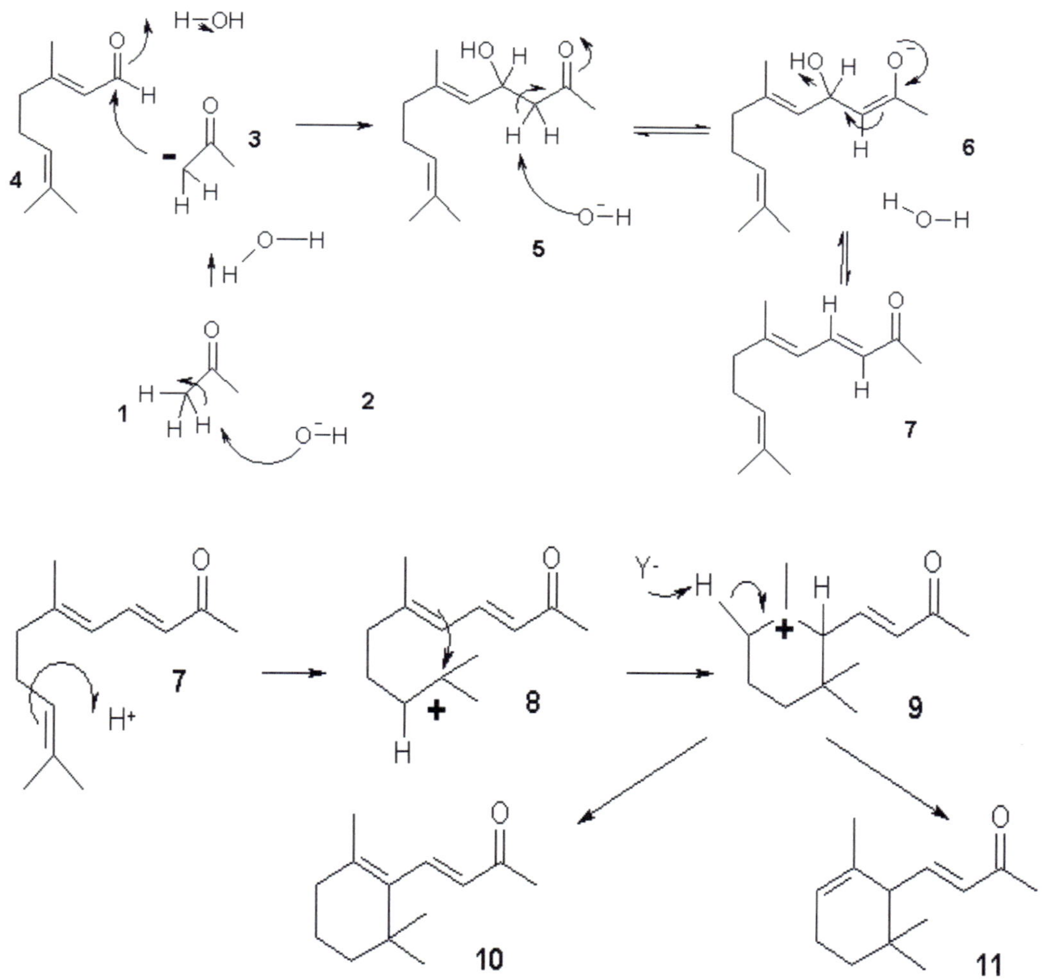

Scheme 3. Synthesis of ionones pseudoionones from citral.

1.6.4. *Synthesis of vitamin E (tocopherol)*

Vitamin E can be synthesized from pseudoionones and tri-methyl hydroquinone (TMHQ) as shown in **Scheme 6**.

1.7. Citral acetals and their importance in cosmetics and toiletries

Citral has strong lemon-like aroma but it is highly volatile and instable to sunrays and alkalis, thus hardly sustaining its aroma [11]. To resolve this problem, citral dimethyl acetal and citral diethyl acetal have been used these compound have neroli-like citrus green

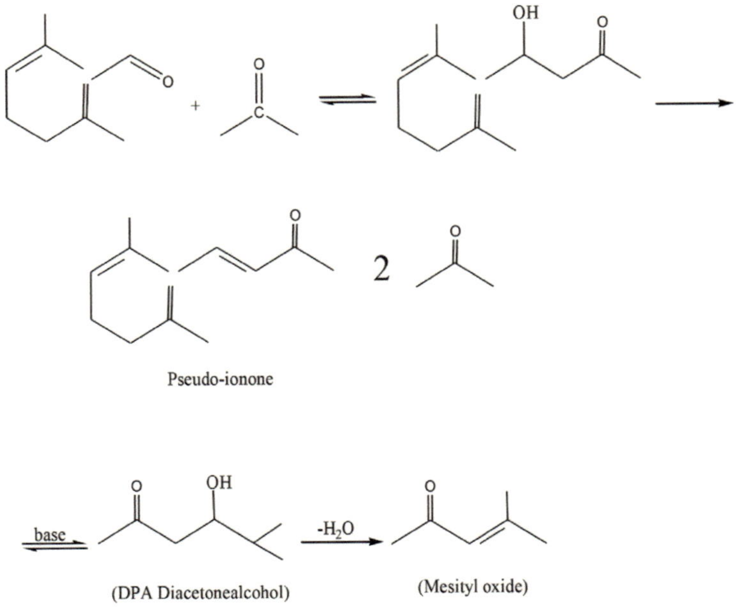

Scheme 4. Condensation of citral.

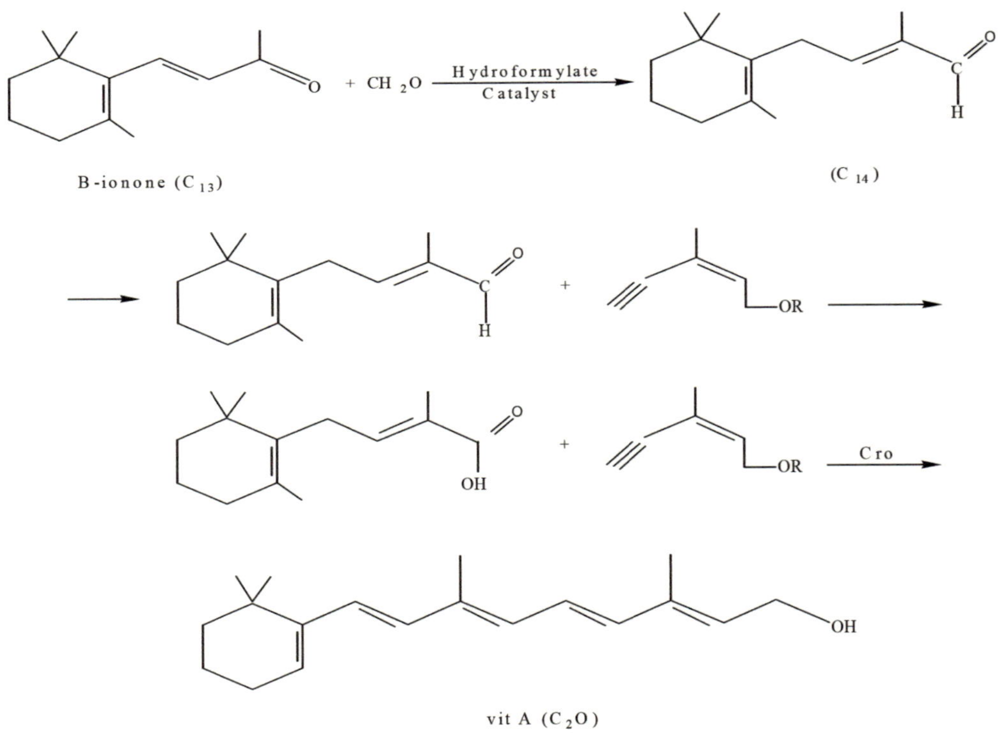

Scheme 5. Catalytic isomerization and synthesis of vitamin A.

http://dx.doi.org/10.5772/intechopen.69518

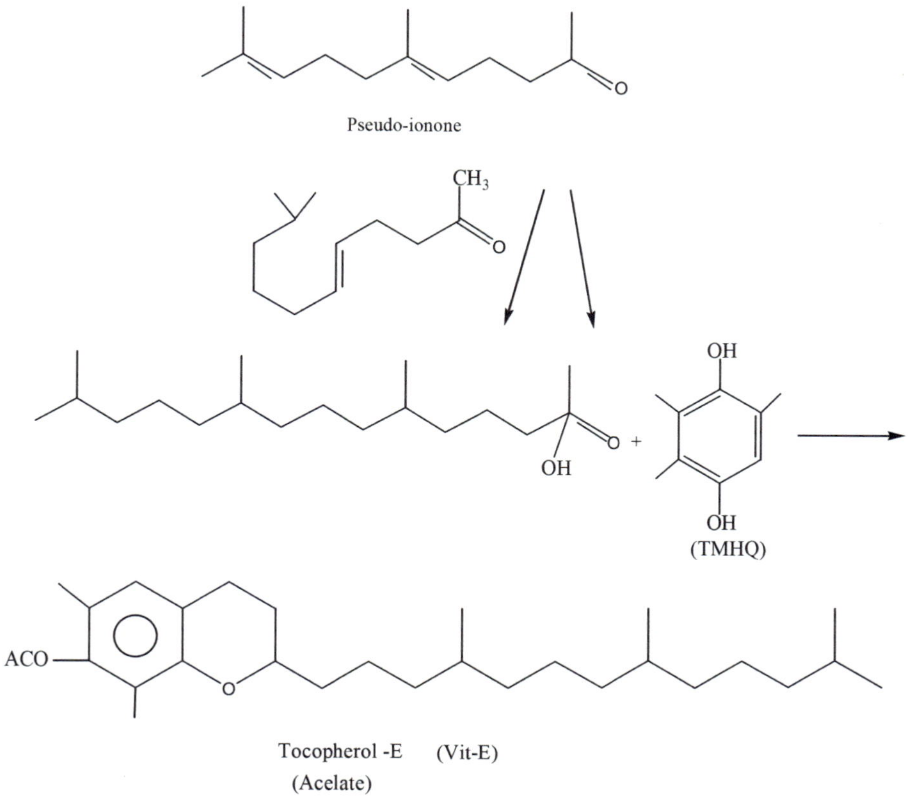

Scheme 6. Synthesis of vitamin E.

aromas. Citral propylene acetals and mono-ether glyceryl acetals are also synthesized and used in cosmetic.

These citral acetals are used in cosmetics and toiletries to suppress body smell produced by bacteria instead of antibacterial agents. Citral acetals in cosmetic formulation slowly release citral, which is antibacterial and also relative save to use.

1.8. Synthesis of citral acetals

When aldehydes or ketones are treated with alcohols in the presence of acids, it produces an acetal sometimes called ketal [12, 13]. The acetal group has two OR groups attached to the same carbon atom. The formula of acetals is not favored when ketones are treated with simple alcohols and gaseous hydrogen chloride. The formation of cyclic acetals is favored with an excess of 1,2-diol and a trace of acid. The reaction can be reversed by treating the acetal with aqueous acid. Citral ethylene glycol and propylene glycol acetals produced by azeotropic acid catalyzed reaction as shown in **Scheme 7**.

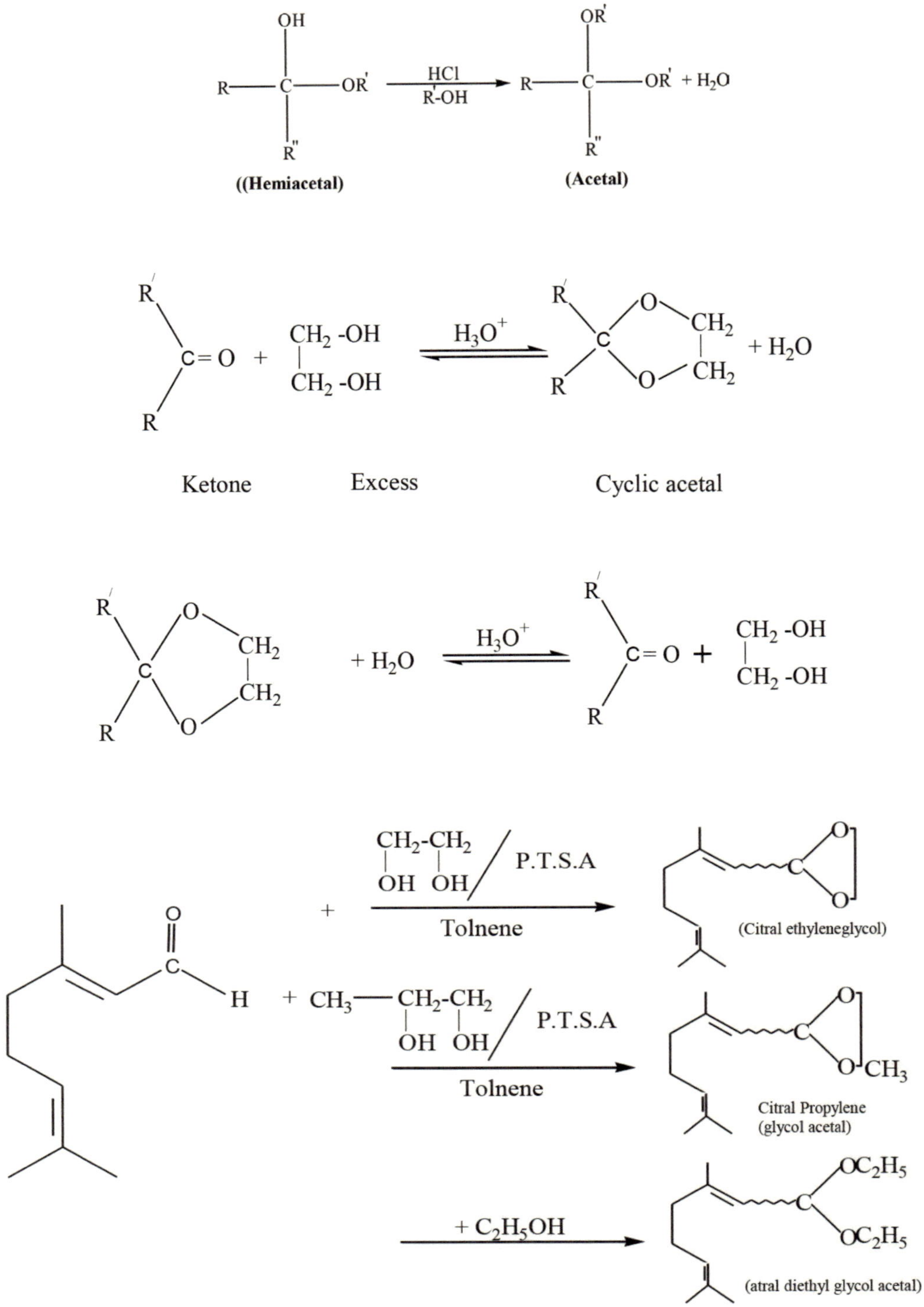

Scheme 7. Synthesis of citral acetal.

2. Material and methods

Lemon grass leaves were harvested in the month of July, allowed to dry at room temperature for 2 days and cut into pieces of 2 inches in length then distilled in an appropriate stainless steel equipment having a capacity of 15 L along with a condenser having inlet and outlet for water, a tank is heated with a gas burner to continue and complete distillation. **Figure 2** shows a stainless steel tank for distillation.

The azeotropic mixture of essential oil is collected in a separating funnel and the essential oil of lemon grass (5.7 g) was obtained, dried over anhydrous sodium sulfate, and kept in an air tight stoppered bottle. The percentage yield of oil was determined by infrared spectroscopy (IR) and gas chromatography (GC), and obtained data lemon grass oil were plotted on graphs.

2.1. Hydrodistillation of essential oil of lemon grass in Dean-Stark apparatus

Citral (lemon grass oil) was hydrodistilled by using Dean-Stark apparatus, which consists of a round bottomed flask with the capacity of 2 L. Three-fourth of the flask was filled with 500 g of crushed dried leaves of lemon grass along with water and hydrodistilled. The oil was separated and dried over anhydrous sodium sulfate. It yields 0.38% oil. **Figure 3** shows Dean-Stark apparatus.

2.2. Analytical equipment used

Infrared spectrophotometer: Model Thermo Nicolet FT-IR 200 (USA) was used for recording absorption in the infrared region.

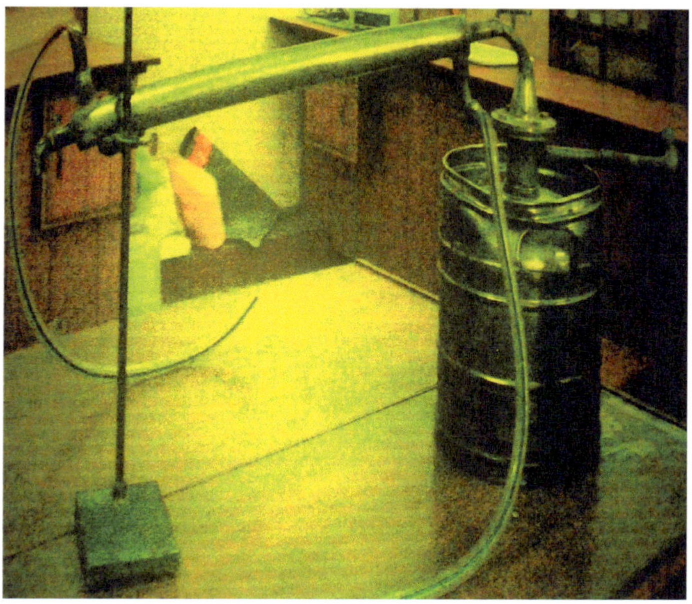

Figure 2. Stainless steel equipment for hydrodistillation of lemon grass.

Figure 3. Dean-Stark apparatus.

Gas chromatograms: Gas chromatography studies were carried out on "Shmadzu" Model GC 14A, packed column SE-30, stationary phase PEG (polyethylene glycol), temperature condition 80–210°C, 5°C increment/min sampling recording temperature 80–210°C/10 min.

- GC of citral ethylene glycol acetal was carried out on packed column SE-30, column condition 185°C° for 2 min to 210°C for 5 min, detector flame ionization detector (FID) 270°C, injector 240°C.

2.3. Citral acetals

2.3.1. Distillation of citral

The purpose of experiment is to synthesize citral acetals, citral was redistilled under vacuum and collected the fraction between 110 and 117°C. Then IR and GC of this citral were recorded and graphs were plotted.

2.3.2. Preparation of citral propylene glycol acetal

In a round bottomed distillation flask (500 ml), a solution of citral (16 g), propylene glycol (22 g) in toluene (60 ml), and 5–7 crystals of para-toluene sulfonic acid as a catalyst is added. Then fit the flask with Dean-Stark apparatus and heated the flask at 110°C with continuous stirring. An azeotropic mixture of water and toluene was distilled off, sodium bicarbonate used to neutralize the reaction residue, benzene is used for extraction of above residue and it is dried over $MgSO_4$ then benzene distilled off and the residual product was redistilled under vacuum and product fraction at 121–125°C is citral propylene glycol having fruity odor. GC and IR of the fractions were recorded.

2.3.3. Citral ethylene glycol acetal

To a citral solution (8 g) was added ethylene glycol (11 g) and benzene (10 ml) with few crystals of para-toluene sulfonic acid in a 250 ml round bottomed flask and stirred by heading. Water was expelled as azeotrop blend with benzene. The remaining item was extracted with hexane and cleaned with sodium bicarbonate arrangement, washed with water and hexane layer was passed over magnesium carbonate. The hexane concentrate was refined off. The left in the cup was refined under vacuum, utilized and gathered three parts at various divisions. These fractions gave fruity odor. GC and IR are recorded.

2.3.4. IR absorption spectroscopy of compounds

A very small amount of the compounds lemon grass essential oil, distilled citral, citral propylene glycol acetal and citral ethylene glycol acetal), was placed separately between two high purity plates of sodium chloride with the help of hypodermic syringe. IR spectra were then taken for these liquids. Before sampling the plates were washed with anhydrous ether then the compounds are smeared between two plates and spectra were recorded.

2.3.5. GC of compounds

A very small amount of compounds (citral and citral acetals) was injected in to a column of PEG-coated on celite support. Nitrogen was used as the carrier gas the flame ionization detector was fitted with it, the temperature was kept at 80–210°C. The peaks obtained were then identified and results were noted.

3. Results

Essential oil from lemon grass was extracted by hydrodistillation. This oil, also known as citral, is tested for its chemical composition and functional groups by gas chromatography-mass spectrometry (GC-MS) and IR spectroscopy and obtained the results.

Citral acetals by using citral was synthesized and tested by GC and IR spectroscopy, results are obtained and graphs are plotted.

3.1. Interpretation of IR spectra of lemon grass

The IR spectra of lemon grass oil having strong characteristic peaks at 3476 show the presence of OH and peaks at 2967, 2917, 2856, and 2759 cm^{-1} show the C–H stretching, a peak at 1686 shows the unsaturated conjugated C=O group present in citral, and peaks at 1650 1613, and 1445 show the C=C stretching, (see **Table 1** and **Figure 4**).

Sr.#	Peaks	Intensity	Assignment
1	3476	Broad	–OH
2	2967	Sharp & Str	–C–H
3	2917	Sharp & Str	C=O
4	2856	Sharp	CH_3, CH_2, C–H
5	1710	Str	C=O
6	1686	Sharp & Str	α,β unsaturation
7	1650	Sharp	–HC=CH–
8	1613	Sharp	–HC=CH–
9	1445	Sharp	C–H
10	1376	Sharp & Str	O–H
11	1194	Sharp & Str	$R—CH_3—\overset{O}{\overset{\Vert}{C}}—OR$
12	1153	Str	–HC=CH–
13	1120		
14	1043	Str	$R—O—\overset{O}{\overset{\Vert}{C}}—$

Table 1. IR absorption spectra of lemon grass oil.

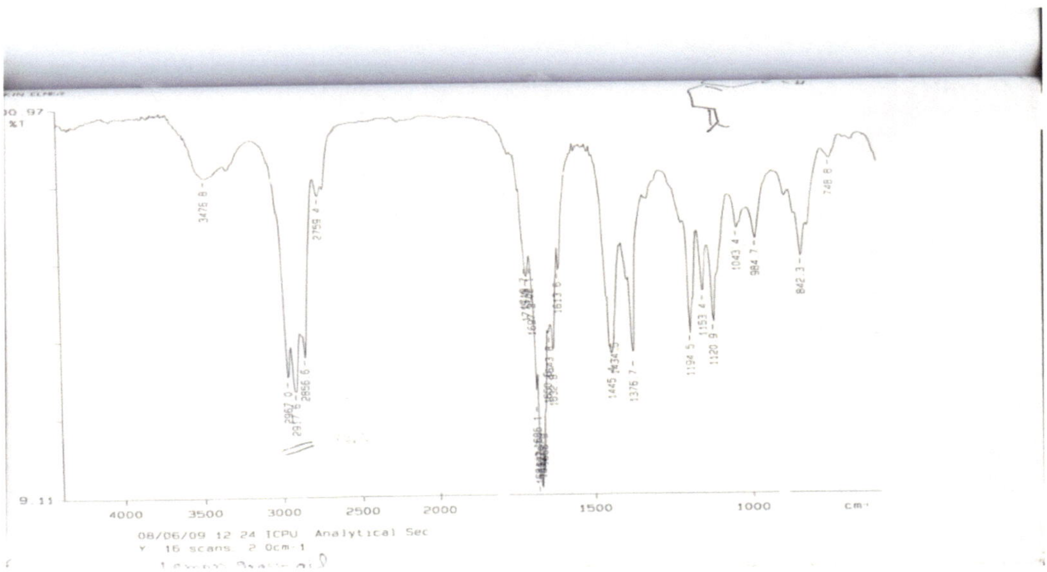

Figure 4. IR absorption spectra of lemon grass oil.

3.2. Interpretation of GC data of lemon grass oil

Different fractions of lemon grass oil spectra on GC showed peaks of six major components. It contained 36, 40 and 34% citral A and 25, 20 and28% citral B, and 7, 5 and 8% in three fractions of steam distilled essential oils, respectively. Fractions 1, 2, and 3 also showed peaks of myrcene along with geranial and other isomers. It is reported that essential oil is usually contained citral A (47%), citral B (33%), myrcene (10%), and geraniol (2%) by GC analysis, see **Table 2** and **Figure 5**, obtained by gas chromatography and gas chromatograph of lemon grass.

Sr#	Compound name	Percentage composition		
		F1	F2	F3
1	Citral (cis and trans)	36.5	40	34.6
2		25.2	10	16.4
3	Myrcene	7.1	10	11.6
4	Other isomers	3.7	5.4	7.4
5		3.7	4.8	5.4
6		3.4	4.4	4.5
7	Geraniol	2.6	3.8	3.0
8	Other isomers	2.2	3.8	
9		1.9	2.9	
10		1.1	1.9	
11		0.9	1.3	
12		0.9	1.3	
13		0.8		
14		0.8		
	Total	86.2	89.6	82.9

Table 2. GC data for lemon grass oil.

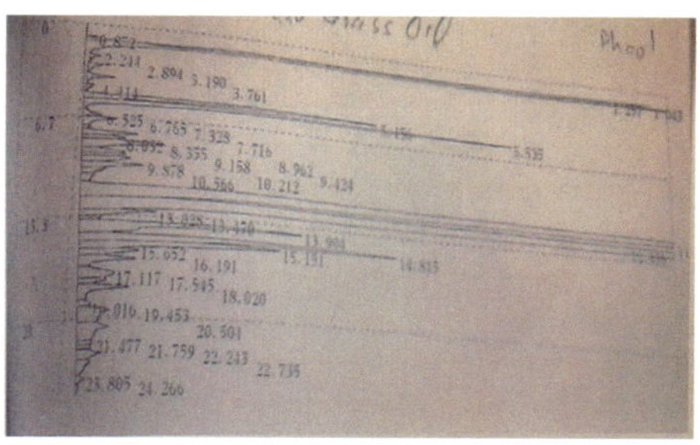

Figure 5. GC data for lemon grass oil.

3.3. Interpretation of IR spectra of citral-propylene glycol acetal

The IR spectra indicate peaks at 3019, 2925, and 2869, which is associated with the C–H stretching. The medium top peaks at 1667, 1514, 1460, and 1380 show the presence of the C=C stretching due to unimmersion in citral part in this compound. The extending vibrations at 1056, 1106, are because of ether linkage in the given compound (see **Table 3** and **Figure 6**).

3.4. Translation of GC information of citral propylene glycol acetal

The response blend of citral propylene glycol acetal later extraction with benzene, redistilled under vacuum at 64–88°C. Noteworthy, parts are appeared in **Table 4**. The redistilled compound is thymol rose. As the citral contained cis and trans geometrical isomers, it gave numerous items by response to propylene glycol, because of taking after reason.

The citral propylene glycol acetal anticipated that it would give four more isomers because of development of two hilter kilter focuses at C2 and C4 of 1,3-dioxo-4 methyl-citral acetal (see **Table 4** and **Figure 7**).

Sr.#	Peaks			Intensity	Assignment
	F1	F2	F3		
1	3019	3019	3019	Sharp & Str v. close	–CH_3,CH_2, C–H
2	2925	2925	2925	Sharp & Str	C=O
3	2871	2871	2869	Sharp	CH_3,CH_2,C–H
4			1894	Wk & sharp	Isomers
5			1744	Wk & sharp	
6			1667	Wk & sharp	–CH=CH–
7	1608	1608		Med	Unsaturation
8	1514	1514	1514	Sharp & med	C=C
9	1460	1460	1460	Sharp & Str	O–H
10	1380	1380	13880	Sharp & Str	R—CH_3—C(=O)—OR
11	1108	1106	1106	Str	
12	1055	1055	1056		R—O—C(=O)—

Table 3. IR absorption data of citral propylene glycol acetal.

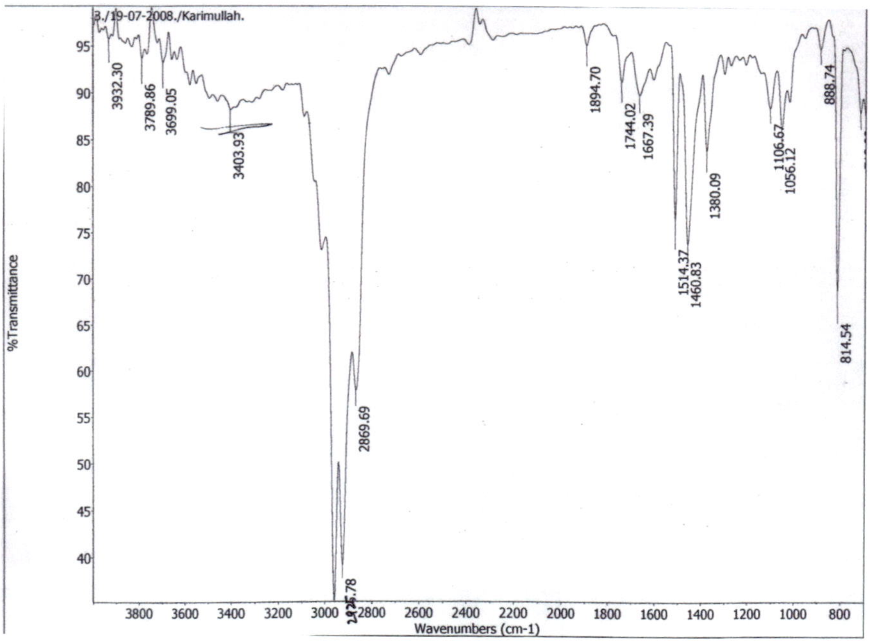

Figure 6. IR absorption spectra of citral propylene glycol acetal.

Sr.#	Compound name	% Age composition F1	F2
1	Citral propylene glycol acetal	55.4	31.2
2	Citral (cis/trans)	16.23	18.5
3		6.5	10.3
4		6.1	6.2
5	Other isomers	4.9	5.8
6		4.5	4.2
7		2.9	4.0
8		1.2	3.5
9		0.7	2.4
10		0.4	2.1
	Total	98.83	95.7

Table 4. GC data of citral propylene glycol acetal.

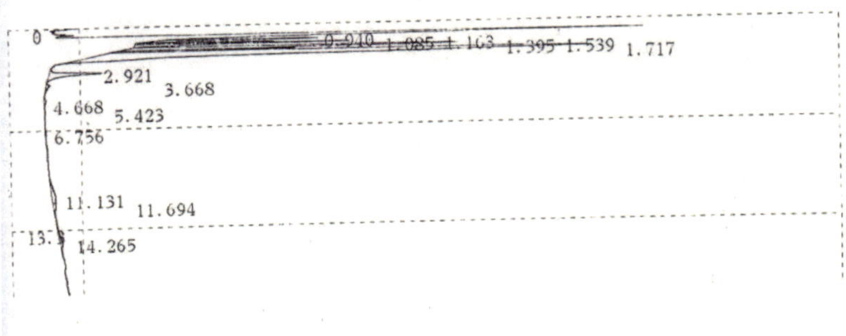

Figure 7. GC data for citral propylene glycol acetal.

3.5. Interpretation of infrared absorption spectra of citral ethylene glycol acetal

The IR spectra of citral ethylene glycol acetal have trademark top at 2924, which demonstrates the C–H extending. Top at 2867 demonstrates the nearness of CH_3, CH_2, and CH in the given compound. Crest at 1665 shows unimmersion in compound. Top at 1055 demonstrates the nearness of C–O and at 814 shows C=C. Other crest at 1513, 1460, and 1379 demonstrates the nearness of different isomers in the compound (see **Table 5** and **Figure 8**).

Sr.#	Peaks	Intensity	Assignment
1	3747	Wk	
2	3649	Med	C–H
3	2924	Sharp & Str	C–H
4	2867	Sharp & Str	CH_3,CH_2,CH
5	1665	Med	C=C
6	1513	Sharp & Str	C=O
7	1460	Sharp	C–H
8	1379	Med	OH
9	1055	Med	$R—CH_3—C(=O)—OR$
10	814	Med	C=C

Table 5. IR absorption data of citral ethylene glycol acetal.

http://dx.doi.org/10.5772/intechopen.69518

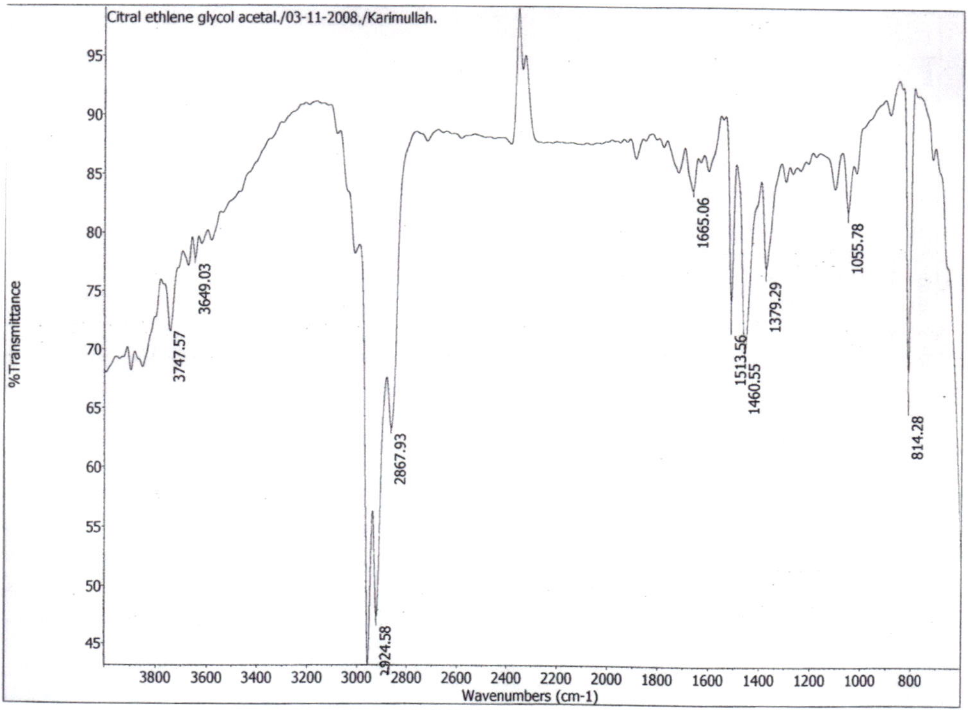

Figure 8. IR absorption spectra of citral ethylene glycol acetal.

3.6. Elucidation of GC information of citral ethylene glycol acetal

Citral ethylene glycol acetal was subjected to GC examination and rate arrangement of major components was resolved by GC, indicating two pinnacles of citral ethylene glycol acetal which are two isomers: cis and trans. This is defended as the beginning citral has two geometrical isomers. Different isomers are additionally present in little sum (see **Table 6** and **Figure 9**).

Sr#	Compound name	% Age composition
1	Citral ethylene glycol acetal	52.7
2	(Cis & trans)	38.5
3	Other isomers	8.3
4		0.3
Total		99.8

Table 6. GC data for citral ethylene glycol acetal.

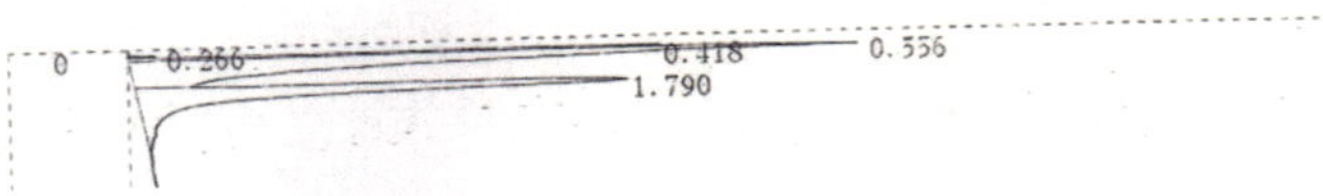

Figure 9. GC data for citral ethylene glycol acetal.

4. Conclusion

Citral is a vital transitional synthetic for the blend of flavors and vitamin A and vitamin E.

Citral is acquired from fragmentary refining of lemon grass too add up to blend from myrcene and 2-methy-4-hydroxy-yet 1-ene on business scale. The citral acetals are additionally imperative halfway for the union of flavor and in addition to their utilization in aroma and beauty care products.

The R&D development of the syntheses of various acetals and ketals is being carried out recently and their biochemical aspects are also under investigations.

Author details

Miss Phool Shahzadi

Address all correspondence to: psk_a@yahoo.com

GCRC, PCSIR, Labs. Complex Lahore, Pakistan

References

[1] Onawaunmi GO. Evaluation of the antimicrobial activity of citral. Letters in Applied Microbiology. 1989;**9**(3):105-108

[2] Robacker DC, Hendry LB. Neral and geranial components of the sex pheromone of the parasitic wasp, Itopletis conquisitor. Journal of Chemical Ecology. 1997;**3**(5):563-577

[3] Mercier C, Chabardes P. Organometallic chemistry in industrial vitamin A and vitamin E synthesis. Pure and Applied Chemistry. 1994;**66**(7):1509-1518

[4] Abegaz, B. and Yohannes, P.G. (1983). Constituents of the essential oil of Ethiopian Cymbopogon citratus Stapf. J. Natural Products. **46**:424-426. doi: 10.1021/np50027a022

[5] Fenaroli's Handbook of Flavor Ingredients, 5th ed - CRC Press - P. 1864 - 1950 B/W Illustrations ISBN 9780849330346 - CAT# 3034

[6] Aromatic Plant Project. San Francisco, CA 94117

[7] LIN, T.S. (1981) Study on the variation of yield and composition of essential oil from *Litsea cubeba*. Bulletin of the Taiwan Forestry Research Institute, No. 355. 14 pp.

[8] Robbins, S. R. J, Citrohella, Lemon Grass & Eunclyptus Oils. London: Tropical Production Institution (TPI); 1983

[9] Shahdab Q, Hanif M, Chandhary FM. Antifungal activity by lemon grass essential oils. Pakistan Journal of Scientific and Industrial Research. 1992;**35:**246-249

[10] Sabcho Dimitrov, Yana Koleva, T. Wayne Schultz, et al. (2004), Enviromental Toxicology and chemistry, Interspecies quantitative structure-activity relationship model for aldehydes: Aquatic toxicity, DOI: 10.1897/02-579

[11] Gradeft P. as Patent 3840601 (1975). Constamitino V, Apperly DC. Catalysis Letters. 1994;23:361 p

[12] Kamogawa, et al. Chemical Society of Japan. 1981;54(**5**):1577-1578 (Accession Number **1981**: 497641 CAPLUS).

[13] Shahzadi P, Alam S, Muhammad A. Book on Synthesis of Citral Acetals. Saarbrücken, Germany: Lap, OmniScriptum GmbH & Co. KG; 2014. p. 64. ISBN-13,9783659636110

Chapter 8

Vetiver Grass: A Tool for Sustainable Agriculture

Suarau O. Oshunsanya and OrevaOghene Aliku

Additional information is available at the end of the chapter

http://dx.doi.org/10.5772/intechopen.69303

Abstract

Vetiver grass is a densely tufted bunch grass which can be easily established in both tropics and temperate regions of the world. It plays a vital role in watershed protection by slowing down and spreading runoff harmlessly on the farmland, recharging ground water, reducing siltation of drainage systems and water bodies, reducing agro-chemicals loading into water bodies and for rehabilitation of degraded soils. Vetiver grass could tolerate extremely high levels of heavy metals. It could be used as biological pest control. The use of vetiver grass has been regarded as a low-cost technology for soil and water conservation; on- and off-farm land and water sources stabilization and remediation of polluted soils; and enhancement of water quality for irrigation purposes when compared with other soil conservation technologies. It could be a dynamic tool for mitigating environmental and agricultural problems, thereby enhancing crop yield and supporting all-year round agricultural cultivation. Recently, vetiver grass has been used to raise animals of different kinds. Thus, this chapter in the book explores several applications of vetiver grass, its impacts and resultant benefits as a technology that could enhance sustainable agricultural development.

Keywords: soil conservation, bioremediation, soil amendment, crop productivity, soil stabilization

1. Introduction

Soil is an integral part of the environment that is involved in many ecosystem services. However, decline in the actual and/or potential productivity of soils due to poor land management practices has become a major challenge to sustainable agriculture and environmental quality [1], thus threatening the food security of many countries of the world. According to Truong [2], some of these poor land management practices often lead to soil erosion and agro-chemical contamination from agricultural practices, urban wastes and industrial operations, which adversely reduces soil's potential for sustainable food production, consequently affecting plants, animals and human lives.

In Truong [2], soil conservation methods such as contour banks, earthen bunds and concrete structures employed to rehabilitate lands with low soil quality are expensive, short-lived and sometimes impossible to execute. He finally reported that vegetative methods, such as the use of grasses, are an effective erosion and sediment control technology which are practicable and economical for the rehabilitation of lands for agricultural production. Vetiver grass (*Vetiveria zizanioides* L.) has been reported to be effective in erosion control [3] and remediation of contaminated soils [4]. According to Rao [5], due to their efficiency and low-cost, vetiver systems are more profitable than both engineering structures and other vegetative barriers. In Ref. [6], vetiver was reported to be highly tolerant to extreme soil conditions including heavy metal contaminations. In comparison with other grasses for rehabilitation of lands with adverse soil conditions, vetiver grass was reported to be superior to Bermuda grass (*Cynodon dactylon*) which has been recommended as a suitable species for acid mine rehabilitation [7]. Also, in an attempt to revegetate a highly saline land, Truong [8] reported superior performance of vetiver grass over Rhodes (*Chloris guyana*) and saltwater couch (*Paspalum vaginatum*), as it was able to survive and resume growth under very high saline conditions. This is due to its unique morphological and physiological characteristics which enables it to survive where other plants cannot [9].

2. Brief description of vetiver grass

Vetiver grass (*V. zizanioides* L.) is a perennial tufted plant that is native to India (**Figure 1**). It is described as a coarse Asian plant [10]. Although *V. zizanioides* is also commonly found in West Africa, an African counterpart, *V. nigritana*, has been reported to be far more common [10]. **Table 1** shows some countries where vetiver grass is known to exist. According to Truong [11], vetiver grass possesses a root system that is abundant, complex, extensive and vertical in nature. According to National Research Council [10], the root grows almost straight down with few lateral surface roots, thus not interfering with the growth of other crops which could result in loss of yield. In Hengchaovanich [12], it was explained that the root system of vetiver grass can reach 3–4 m in the first year of planting, while [13] reported that it attains a total length of 7 m after 36 months. The roots are very strong with a mean tensile strength of between 75 and 85 MPa [14]. The leaves are thin and have sharp edges [10], while the shoots can grow up to 2 m. The mature foliage is tough and coarse, which enables it to stay in place for years [10]. They further reported that this attribute is important as an erosion control crop provided it is to work effectively. Its growth occurs from the crown, which rises relative to soil build-up [9]. The crown of the plant occurs slightly below the soil surface and as such no lasting damage can be done on it by grazing or trampling animals [10].

2.1. Vetiver grass technology

Vetiver grass has been reported to be very effective in trapping both fine and coarse sediments in runoff water [10]. These sediments constitute bulk of the fertile layer (topsoils) of

Figure 1. Vetiver (*Vetiveria zizanioides* L.) grass strips at the University of Ibadan, Nigeria.

Africa	America	Asia	Caribbean	Pacific	Others
Algeria	Argentina	Bangladesh	Antigua	American Samoa	France
Angola	Brazil	Burma	Barbados	Cook Islands	Italy
Burundi	Colombia	China	Cuba	Fiji	Spain
Comoro	Costa Rica	India	DR	New Caledonia	USA
CAR	FG	Indonesia	Haiti	New Guinea	USSR
Ethiopia	Guatemala	Japan	Jamaica	Tonga	
Gabon	Guyana	Malaysia	Martinique	Western Samoa	
Ghana	Honduras	Nepal	Puerto Rico		
Kenya	Paraguay	Pakistan	St. Lucia		
Madagascar	Suriname	Philippines	St. Vincent		
Malawi		Singapore	Trinidad		
Mauritius		Sri Lanka	Virgin Islands		
Nigeria		Thailand			
Rwanda					
Reunion					
Seychelles					
Somalia					

Africa	America	Asia	Caribbean	Pacific	Others
South Africa					
Tanzania					
Tunisia					
Uganda					
Zaire					
Zambia					
Zimbabwe					

Note: CAR, Central African Republic; FG, French Guiana; DR, Dominican Republic (source: Ref. [10]).

Table 1. Some countries where vetiver is known to exist.

most agricultural lands, which is critical for crop cultivation. In addition, vetiver grass has been reported to have high tolerance for extreme adverse conditions, including heavy metal toxicity [15], hence making it suitable for the remediation of heavy metal-contaminated soils. This could be due to some of its special attributes which makes it an ideal species for environmental protection and sustainable agriculture. Some of these characteristics include massive, fine-structured root system [16]; high resistance to pests, diseases and fire [17]; high efficiency in absorbing dissolved N, P, Hg, Cd and Pb in polluted water [18]; and good and fast recovery rate after being affected by the previously listed adverse conditions [19].

2.2. Applications of vetiver grass

The application of vetiver grass as a technology for soil and water conservation was first developed in India by the World Bank in the 1980s [2]. Some of the applications of vetiver grass technology, which could sum up to the enhancement of sustainable agricultural development, include its use for soil erosion and sediment control on sloping farmlands and floodplains [20–24]; rehabilitation of saline and acid sulphate soils [6]; bioremediation of agrochemicals [25, 26]; biological pest control [24]; and on- and off-site heavy metal pollution control [6, 26] amongst others.

2.2.1. Erosion and sediment control

Erosion, which is simply the washing away of soils by 'agents' such as water and wind, is a phenomenon that has ravaged so many lands, resulting in soil degradation and consequently low crop yield. According to National Research Council [10], it is among the most devastating environmental disaster for many developing countries and it results in loss of huge amounts of valuable soils which are key to agricultural production. Management methods could be expensive and sometimes less effective. However, Truong [27] reported that both research and field results in Australia, Asia, Africa and South America show that in comparison with conventional cultivation practices, surface runoff and soil loss from fields treated with vetiver grass were significantly lower and crop yield was much improved. **Figure 2** illustrates the processes of erosion and sediment control under conventional cultivation practice and vetiver grass system.

http://dx.doi.org/10.5772/intechopen.69303

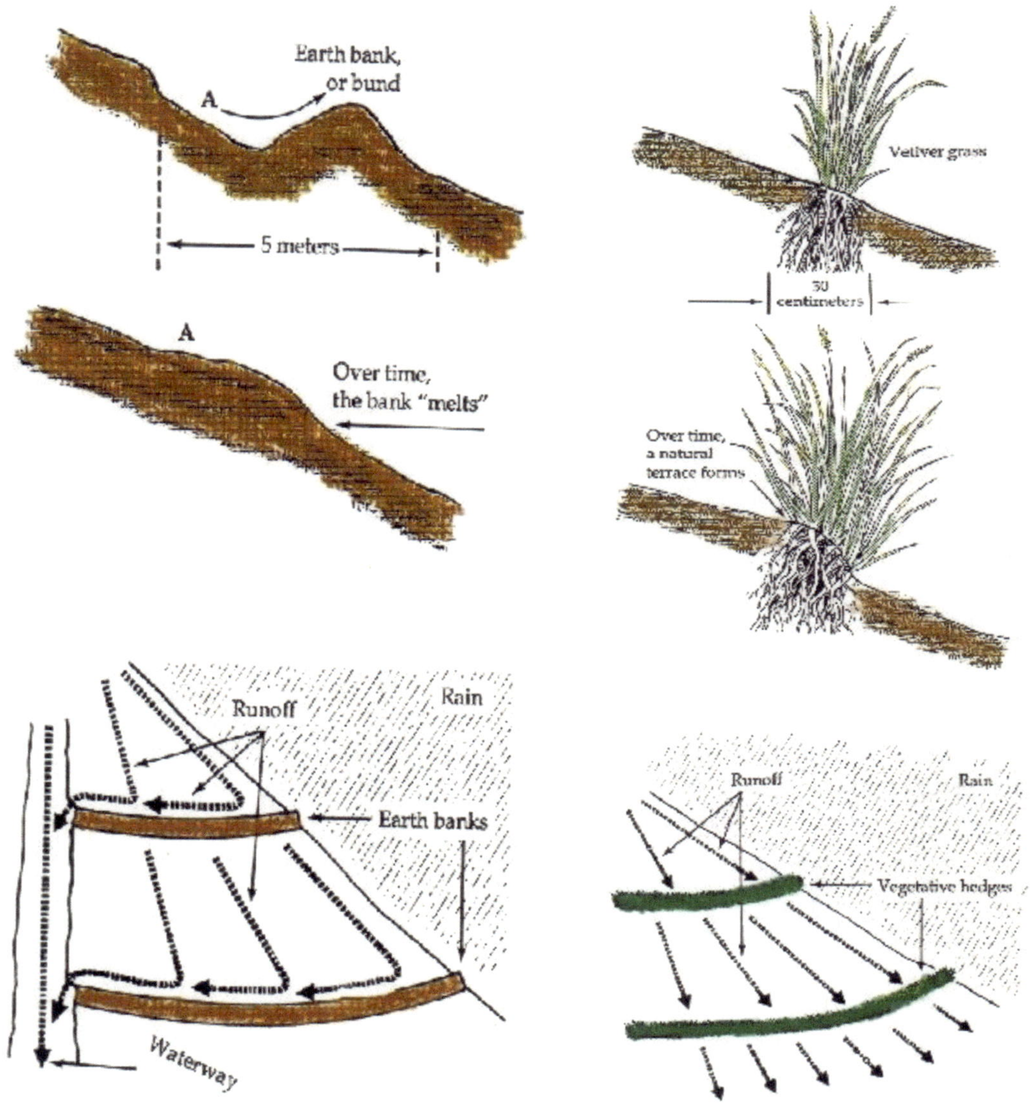

Figure 2. Comparison between conventional terrace/contour system and Vetiver system in soil and water conservation (adapted from Ref. [9]).

V. zizanioides has been reported to reduce soil loss from 11 to 3 t ha^{-1} [10]. They explained that vetiver is suitable for erosion and sediment control because it slows runoff and gives the rainfall a better chance of soaking into the soil instead of rushing off the slope. According to Truong and Loch [9], when vetiver grass is planted in single or multiple rows on the contour, it forms a protective barrier across the slope, which slows the runoff water, thereby causing sediment to be deposited. They further explained that since the barriers only filter the runoff and do not convey it, water seeps through the hedge, reaching the bottom of the slope at lower velocity without causing any erosion or without being concentrated in any particular area.

According to Refs. [21–23, 26, 28–30], results over the last 10 years have showed vetiver grass to be very successful in reducing flood velocity and limiting soil movement, with very little erosion in fallow strips. In Rao et al. [31], relative to control plots, average reductions of 69% in runoff and 76% in soil loss were recorded from vetiver plots. In Nigeria, Babalola et al. [3] demonstrated the usefulness of vetiver grass as a soil and water conservation measure in the Nigerian environment. They established vetiver strips on 6% slopes for three growing seasons. The results of the study showed that vetiver grass ameliorated soil physical and chemical conditions, reduced soil and nutrient losses, and increased soil moisture storage by a range of 1.9–50.1% at various soil depths for a distance of 20 m. Other research studies that have been conducted on the use of vetiver grass for soil and water conservation in Nigeria include Refs. [21–23, 26, 32, 33]. **Table 2** shows a summary of the effects of vetiver grass and conventional systems in soil loss and runoff control.

2.2.2. *Rehabilitation of saline and acid sulphate soils*

Salinity, which is the amount of dissolved salt content of a soil or water body, is a major challenge confronting agricultural production especially in semi-arid regions of the world with respect to crop. According to Truong and Baker [35], vetiver grass that could cope with saline soils has been successfully employed in the rehabilitation of salt-affected lands. This may be due to its high tolerance to salt-affected soils. In Truong [36], it was reported that with a salinity threshold level at $EC_{se} = 8$ d Sm^{-1}, vetiver grass compares favourably with some of the most salt-tolerant crops and pasture species (such as Bermuda grass (*C. dactylon*) with threshold at 6.9 d Sm^{-1}; Rhodes grass (*C. guyana*) at 7.0 d Sm^{-1}; Wheat grass (*Thynopyron elongatum*) at 7.5 d Sm^{-1}; and barley (*Hordeum vulgare*) at 7.7 d Sm^{-1}) grown in Australia.

Country	**Soil loss**			**Runoff**		
	Control	**Conventional**	**Vetiver system**	**Control**	**Conventional**	**Vetiver system**
Thailand	3.9	7.3	2.5	1.2	1.4	0.8
Venezuela	95.0	88.7	20.2	64.1	50.0	21.9
Venezuela (15%)*	16.8	12.0	1.1	88	76	72
Venezuela (26%)*	35.5	16.1	4.9	NA	NA	NA
Vietnam	27.1	5.7	0.8	NA	NA	NA
Bangladesh	NA	42	6–11	NA	NA	NA
India	NA	25	2	NA	NA	NA
		14.4	3.9		23.3	15.5
Nigeria (Flat)	NA	NA	1.25–10.3	5.3–15.7	NA	1.15–4.07
Nigeria (Mound)	NA	NA	0.68–5.00	15.7–26.5	NA	0.72–4.85

Note: NA, not available (source: Ref. [34]).
*Land slope.

Table 2. Comparison of soil loss and runoff control under conventional and Vetiver grass systems.

On the other hand, acid soils which constitute a major part of arable lands in Africa and Asia are highly erodible and difficult to stabilize and rehabilitate [27]. However, vetiver has been successfully used to stabilize and rehabilitate a highly erodible acid sulphate soil, where the actual soil pH was about 3.5 and oxidized pH was as low as 2.8 [4, 37]. When planted on saline and/or acid sulphate soils, vetiver grass can effectively absorb plant available sodium and metals that contribute to soil salinity or acidity, thereby rehabilitating these soils [35]. Also, the tensile strength of its root system can also contribute to increasing soil strength against runoff and erosion, thus stabilizing the soil slope.

2.2.3. *Bioremediation of agro-chemicals*

Agro-chemicals (pesticides, herbicides, and even fertilizers) have been reported to adversely affect soil properties and water quality. This adverse effect is finally expressed in the quality of crop produced. According to Truong [27], vetiver has played an important role in the decontamination of agro-chemicals due to its ability to retain them within its system, thereby preventing them from contaminating and accumulating in soils and crops. **Table 3** shows the threshold levels of heavy metals to vetiver grass. According to Refs. [18, 25], research conducted in cabbage crops grown on steep slope (60%) in Thailand indicated that vetiver hedges had an important role in the process of captivity and decontamination of agro-chemicals, especially pesticides such as carbofuran, monocrotophos and anachlor. According to Truong et al. [15], planting vetiver grass across drainage lines could serve as a living filter for capturing unwanted foreign chemicals or contaminants before they reach non-polluted soil and downstream areas.

2.2.4. *Biological pest control*

Insects and pests are two of the most destructive biological pests known to cause damage to agricultural crops and consequently leading to reduction in crop yield. The use of agro-chemicals in

Heavy metals	Threshold levels in soil (mg kg^{-1})		Threshold levels in plant (mg kg^{-1})	
	Vetiver	Other plants	Vetiver	Other plants
Arsenic	100–250	2.0	21–72	1–10
Cadmium	20–60	1.5	45–48	5–20
Copper	50–100	NA	13–15	15
Chromium	200–600	NA	5–18	0.02–0.20
Lead	>1500	NA	>78	NA
Mercury	>6	NA	>0.12	NA
Nickel	100	7–10	347	10–30
Selenium	>74	2–14	>11	NA
Zinc	>750	NA	880	NA

Note: NA, not available (source: Ref. [9]).

Table 3. Tolerance levels of vetiver grass and other plants to heavy metals.

controlling most of these pests results in adverse effects on both soils and crops. Results of the research carried out in Guangxi University, China, after subjecting vetiver grass to insect attack, showed that of the 79 species of insect found on vetiver rows, only four attacked young vetiver leaves [24]. He explained that due to few insects that could attack vetiver grass, the damage was minimal. Also, the potential of vetiver extract as a natural pre-emergent weed killer was obtained when methanol extracts of its ground stem and root were found to be very effective in preventing the germination of a number of both monocotyledonous and dicotyledonous weed species [38].

2.3. Management of wastewater

Large volume of waste or contaminated water emanating from industrial or domestic discharges could be difficult or expensive to control, especially when released into the environment. In Truong [36], wastewaters can be managed either by disposal (i.e. total elimination or reduction in volume) or treatment (i.e. improving its quality).

2.3.1. Disposal of wastewater

According to Truong [36], vegetative methods are the only feasible and practicable method available for the disposal of wastewater. Vetiver grass has been reported to be more effective than trees and pasture species in the disposal of domestic and industrial effluent. This is because vetiver grass possesses some unique characteristics suitable for environmental protection purposes [39]. Apart from absorbing toxic elements in wastewater, vetiver grass can also absorb large quantities of water, thus reducing the volume of these waters from the environment. Thus, according to Truong [20], the problem of wastewater disposal can be solved by using such water as irrigation water for growing vetiver grass, where it can be absorbed. Effluent discharge was reduced by sub-surface irrigating vetiver grass rows [15]. This could also reduce potential ponding situations which are not usually favourable for most arable crops. It was reported in Truong and Smeal [40] that in producing a massive growth of biomass (>100 t ha^{-1}), vetiver grass consumes a large volume of water.

According to Truong [36], in quantifying the water use rate of vetiver, a good correlation between water use and dry matter yield of vetiver was obtained. He explained that from this correlation, it was estimated that for a kilogram of dry shoot biomass, vetiver grass would use 6.8 L day^{-1}. Also, Truong and Smeal [40] explained that if the biomass of 12-week-old vetiver, at the peak of its growth cycle, was 40.7 t ha^{-1}, a hectare of vetiver grass would potentially use 279 kL ha^{-1} day^{-1}. According to Truong [36], data from a landfill leachate site showed that vetiver grass can dispose up to 3.8 L m^{-2} day^{-1}. It was reported in Truong and Hart [41] that where other plants such as fast growing tropical grasses and trees, and other crops such as sugar cane and banana have failed, vetiver grass survived. Vetiver grass of about 100 vetiver stands in an area less than 50 m^2 completely dried up the effluent discharge from a toilet block. In addition, Percy et al. [42] reported that 4 and 2 mL of highly contaminated landfill leachate were effectively disposed in a month in summer and in winter by 3.5 ha planting of vetiver, respectively, while Smeal et al. [43] reported that most industries in Queensland are upgrading their treatment process of wastewater by adopting vetiver grass system as a sustainable means of disposing wastewater.

2.3.1.1. *Treatment of wastewater*

Vegetative method is generally the most efficient and common method for water quality improvement [36]. He reported that the attributes of vetiver grass indicate that it is highly suitable for treating polluted wastewater from industries as well as domestic discharges. Some of these attributes include its tolerance to elevated and sometimes toxic levels of salinity, acidity, sodicity, and heavy metals. Vetiver can be used to improve wastewater quality either by absorbing pollutants and heavy metals or by trapping debris, sediment and agro-chemicals in agricultural lands. In Truong [20], it was reported that growing vetiver grass on effluent is one of the effective low-cost technologies of treating wastewater. According to Truong and Hart [41], planting 100 vetiver grass in an area less than 50 m^2 completely dried up effluent discharge from a septic tank. An earlier research by Wagner et al. [44] showed the exceptional ability of vetiver grass in absorbing and to tolerate extreme levels of nutrients, Truong et al. [15] stated that nutrients (N, P, Ca, etc.), herbicides (diuron, trifluralin, fluometuron, etc.) and pesticides (α, β and sulphate endosulfan and chlorpyrifos, parathion and profenofos) could be restrained on site if vetiver grass hedges were established across drainage lines (**Figure 3**).

In China, research showed that vetiver grass can reduce soluble P up to 99% after 3 weeks and 74% of soluble N after 5 weeks [27]. Vetiver grass has also been used to control algal growth. For example, Refs. [45, 46] reported that vetiver grass could remove dissolved nutrients and reduce algal growth within 2 days under experimental condition. In addition, Truong [27] explained that vetiver grass can be used very effectively to control algal growth in water infested with blue-green algae by planting vetiver grass strips at the edges of the streams or in the shallow parts of the lakes where usually high concentrations of soluble N and P occur. The thick culms of vetiver that is just above the soil surface also collected debris

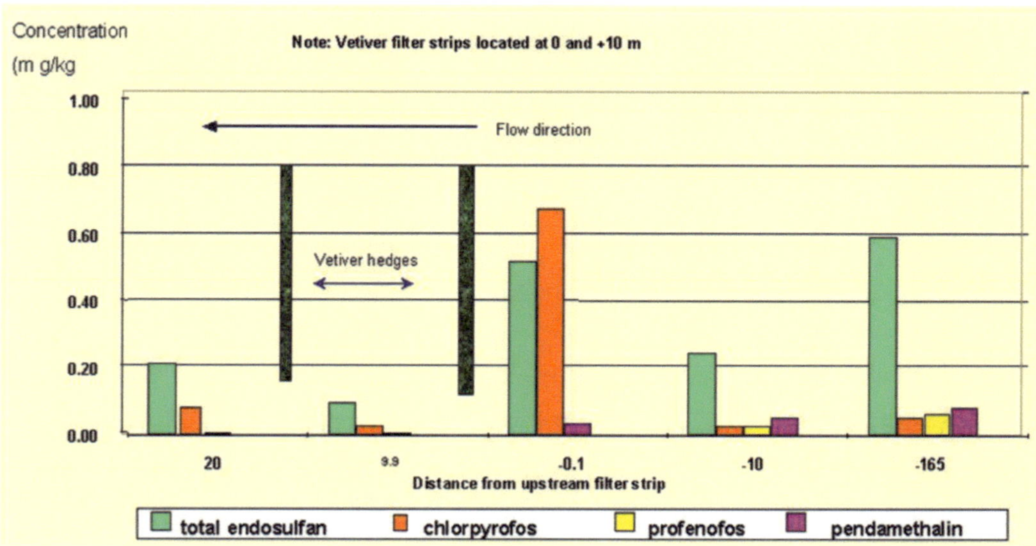

Figure 3. Herbicide concentration in soil-deposited upstream and downstream of vetiver filter strips (source: Ref. [32]).

and soil particles carried along watercourse [47], while Liao et al. [48] reported that with proper planning, vetiver grass technology has the potential of removing up to 102 t of N and 54 t of P yr^{-1} ha^{-1} of vetiver planting.

Furthermore, in the purification of wastewater from a pig farm which contained very high N, P, Cu and Zn, vetiver grass showed a very strong purifying ability, with its ratio of uptake and purification of Cu and Zn (>90%), As and N (>75%), Pb (between 30 and 71%), and P (between 15 and 58%) [48]. Concentrations of some toxic elements (Al, Fe, and SO_4) in vetiver grass planted on an acid sulphate soil was found to increase as the plant matures, thus reducing contamination of canal water [49]. Thus, the efficiency of wastewater treatment increases with an increase in the age of vetiver plant [47]. **Table 4** presents effluent water quality prior to and after vetiver treatment. This could be highly beneficial to agriculture especially in the area of irrigation where available water for crop cultivation is fast diminishing as a result of climate change impact and competition from other users.

2.4. Crop yield

The improvement of agricultural crop yield is one of the resultant benefits of the effects of vetiver grass technology on soil and water conservation. This could be beneficial to farmers, especially those farming on sloppy lands that are usually prone to erosion. It was reported by National Research Council [10] that vetiver grass improves crop harvest by reducing crop failure against the dry spell. They also reported that vetiver grass enhances soil moisture for plant use. In Nigeria, Babalola et al. [3] reported an increase in crop yields by a range of 11–26% for cowpea and by about 50% for maize following the application of vetiver grass strips at 20-m intervals against non-vetiver plots on a 6% slope. They attributed the higher grain yield to higher nutrient use efficiency under vetiver grass strips relative to no vetiver strip.

Also, Oshunsanya et al. [32] reported an increase in maize grain yield (13.5–26.6%), and cassava tuber weight (7.9–11.2%) in a maize/cassava intercrop under vetiver grass strips spaced at 5-, 10- and 20-m surface intervals. Another report by Babalola et al. [50] showed

Parameter	Fresh effluent	Results	
		2002/2003	2004
pH (6.5–8.5)*	7.3–8.0	9.0–10.0	7.6–9.2
Dissolved oxygen (2.0 minimum)*	0–2 mg L^{-1}	12.5–20.0 mg L^{-1}	8.1–9.2 mg L^{-1}
5 day BOD (20–40 mg L^{-1} maximum)*	130–300 mg L^{-1}	29–70 mg L^{-1}	7–11 mg L^{-1}
Suspended solids (30–60 mg L^{-1} maximum)*	200–500 mg L^{-1}	45–140 mg L^{-1}	11–16 mg L^{-1}
Total nitrogen (6.0 mg L^{-1} maximum)*	30–80 mg L^{-1}	13–20 mg L^{-1}	4.1–5.7 mg L^{-1}
Total phosphorus (3.0 mg L^{-1} maximum)*	10–20 mg L^{-1}	4.6–8.8 mg L^{-1}	1.4–3.3 mg L^{-1}

Note: BOD, biological oxygen demand (source: Ref. [32]).
*Licence requirements.

Table 4. Effect of vetiver grass treatment on effluent quality levels.

that grain yields on plots treated with 4 and 6 t ha^{-1} vetiver grass mulch were 4 and 47.4% higher than plots treated with vetiver grass strips, respectively. In addition, Laing [51] finally reported that the full potentials of vetiver grass could be harnessed by combining vetiver grass mulch with vetiver grass strips such that vetiver grass strips would reduce soil water erosion, while mulch materials would decompose to improve the nutrient status of the farmland.

2.5. Other uses of vetiver grass

Apart from its unique physiological characteristics, which give it an edge over other grasses as a plant with diverse environmental applications, vetiver grass has also been found useful in a number of ways. Apart from being a soil and water conservation technology, this grass of great utility has been reported to be legally accepted for use as property lines in certain parts of India. Also, in Nigeria the surveyor general has in past permitted vetiver grass hedges as a legal boundary marker. This is because its bases expand so little [10]. It grows so densely that it can block the spread of weeds. For instance, in Zimbabwe and Mauritius, farmers plant vetiver grass around their fields to keep *Kikuyu* grass and Bermuda grass from invading their fields, respectively [10].

Furthermore, National Research Council [10] reported that for several centuries, vetiver grass has been commercially cultivated for its scented oil that can be distilled from its roots. They also reported that it is a treasured ingredient in some of the world's best-known perfumes and soaps, and largely because of its potential as an export commodity. However, only a handful of countries produce vetiver oil commercially. Although reliable statistics are unavailable, the world production of vetiver oil is estimated to be about 250 tons a year [10]. The annual consumption is estimated in **Table 5**. Other products that could be derived from the vetiver grass include mats, baskets, fans, sachets, window coverings, wall hangings, thatch roofs, lampshades and ornaments which are usually weaved from its roots or stems [10].

Country	Quantity (tons)
United States	100
France	50
Switzerland	30
United Kingdom	20–25
Japan	10
Germany	6
Netherlands	5
Other	30–40

Source: Ref. [10].

Table 5. Estimates of annual consumption of vetiver oil.

3. Conclusions

Vetiver grass technology has been applied globally for controlling soil erosion, stabilizing land and water resources and remediating contaminated lands in order to improve crop growth and yields. It is used as fodder for animal feed, mulch for improving soil moisture and fertility, and fibrous root system for holding soils in place could guarantee food production on a sustainable basis owing to the fact that this grass can withstand adverse environmental and climatic conditions, coupled with quick regeneration after pruning. Thus, when vetiver grass is applied appropriately, it could be a low-cost, simple and easily applicable multi-purpose soil and water conservation tool for sustainable agriculture. It is also a grass of great utility that could provide other means of revenue for local farmers.

Author details

Suarau O. Oshunsanya* and OrevaOghene Aliku

Address all correspondence to: soshunsanya@yahoo.com

Department of Agronomy, University of Ibadan, Ibadan, Nigeria

References

[1] Lal R. Tillage effects on soil degradation, soil resilience, soil quality, and sustainability. Soil and Tillage Research. 1993;**27**(1):1-8

[2] Truong P. Vetiver grass for mine site rehabilitation and reclamation. In: Proceedings of Remade Lands International Conference, Fremantle, Australia; November 2000; pp. 85-86

[3] Babalola O, Jimba JC, Maduakolam O, Dada OA. Use of Vetiver grass for soil and water conservation in Nigeria. In: Proceedings of Third International Conference on Vetiver and Exhibition Guangzhou, China; October 2003; pp. 293-309

[4] Loch R, Truong P, Smirk D, Fulton I. Vetiver grass for land management and reclamation. In: Proceedings of the Third AMEEF Innovation Conference; Brisbane, Australia; August 2000; pp. 116-122

[5] Rao DV. Vetiver Information Network. Newsletter No. 10. ASTAG. Washington D.C.: The World Bank; 1993

[6] Truong PN, Baker, D. Vetiver grass system for environmental protection. Technical Bulletin No. 1998/1998; Pacific Rim Vetiver Network. Office of the Royal Development Projects Board, Bangkok, Thailand

[7] Taylor KW, Ibabuchi IO, Sulford P. Growth and accumulation of forage grasses at various clipping dates on acid mine spoils. Journal of Environmental Science and Health. 1989;**24**:195-204

[8] Truong PN. Vetiver grass for land rehabilitation. In: First International Conference on Vetiver. Bangkok: Office of the Royal Development Projects Board; 1996. pp. 49-56

[9] Truong PNV, Loch R. Vetiver system for erosion and sediment control. In: ISCO 2004—13th International Soil Conservation Organisation Conference; Brisbane; 2004; No. 247. p. 6

[10] National Research Council. Vetiver Grass: A Thin Green Line against Erosion. Washington, D.C.: National Academy Press; 1993. p. 169

[11] Truong PN. Vetiver Grass Technology. In: Maffei, M, editor. "Vetiveria" the Genus Vetiveria. London and New York: Taylor & Francis; 2002. pp. 114-132

[12] Hengchaovanich D. Vetiver grass for slope stabilization and erosion control. Tech. Bulletin' 1998; No. 1998/2, PRVN/Office of the Royal Development Project Board, Bangkok, Thailand

[13] Lavania UC, Lavania S, Vimala Y. Vetiver system ecotechnology for water quality improvement and environmental enhancement. Current Science. 2004;**86**:11-14

[14] Cheng H, Yang X, Liu A, Fu H, Wan M. A study on the performance and mechanism of soil-reinforcement by herb root system. In: Truong P, Xiu HP, editors. Proceedings of the Third International Conference on Vetiver and Exhibition; Guangzhou, China; 2003. pp. 411-414

[15] Truong PN, Mason F, Waters D, Moody P. Application of Vetiver grass technology in off-site pollution control. I. Trapping agrochemicals and nutrients in agricultural lands. In: Proceedings of Second International Vetiver Conference, Thailand; January 2000

[16] Hengchaovanich D. Fifteen years of bioengineering in the wet tropics from A (*Acacia auriculiformis*) to V (*Vetiveria zizanioides*). In: Proceedings of Ground and Water Bioengineering for Erosion Control and Slope Stabilization; Manila; April 1999

[17] West L, Sterling G, Truong PN. Resistance of Vetiver grass to infection by root-knot nematodes (Meloidogyne spp.), 1996. Vetiver Newsletter (20): 20-22.

[18] Sripen S, Komkris T, Techapinyawat S, Chantawat S, Masuthon S, Rungsuk M. Growth potential of Vetiver grass in relation to nutrients in wastewater of Changwat Phetchaburi. Paper presented at ICV-1. In: Abstracts of Papers Presented at First International Conference on Vetiver. Office of the Royal Development Projects Board, Bangkok; 1996. p. 44

[19] Truong P, Baker DE, Christiansen I. Stiffgrass barrier with Vetiver grass. A new approach to erosion and sediment control. In: Proceedings of The Third Annual Conference on Soil & Water Management for Urban Development, Sydney; 1995. pp. 214-222

[20] Truong P. Report on the International Vetiver Grass Field Workshop, Kuala Lumpur. Australian Journal of Soil and Water Conservation. 1993;**6**:23-26

[21] Oshunsanya SO. Surface soil properties and maize yields in runoff plots planted with vetiver grass (*Vetiveria nigritana* Stapf) hedges. Soil Science. 2013;**178**(4):205-213

[22] Oshunsanya SO, Are K, Fagbenro JA. The use of vetiver grass (*Vetiveria nigritana* Stapf) strips in checking soil loss and improving yields of maize-cassava-cowpea inter-cropping systems in Southwest Nigeria. Journal of Applied Agricultural Research. 2014;**6**(1):237-244

[23] Oshunsanya SO. Crop yields as influenced by land preparation methods established within vetiver grass alleys for sustainable agriculture in Southwest Nigeria. Agroecology and Sustainable Food Systems. 2013;**37**(5):578-591

[24] Chen SW. Insect on Vetiver hedges. Assumption University Journal of Technology. 1999;**3**:38-41

[25] Pinthong J, Impithuksa S, Ramlee A. The capability of Vetiver hedgerows on the decontamination of agro chemical residues. In: Proceedings of the First International Conference on Vetiver. Office of the Royal Development Projects Board; Bangkok; 1998. pp. 91-98

[26] Oshunsanya SO, Ewetola EA. Production of hygienic *Amaranthus cruentus* on soils remediated with vetiver grass species in Ibadan, Nigeria. Journal of Science Research. 2011;**10**(2):212-220

[27] Truong PNV. The global impact of vetiver grass technology on the environment. In: Proceedings of the Second International Vetiver Conference, Thailand; 2000. pp. 46-57

[28] Dalton PA, Smith RJ, Truong PNV. Hydraulic characteristics of vetiver hedges: An engineering design approach to flood mitigation on cropped floodplain. In: Proceedings of the First International Vetiver Conference; Chiang Rai, Thailand; October, 1996, 1996a. pp. 65-73

[29] Dalton PA, Smith RJ, Truong PNV. Vetiver grass hedges for erosion control on a cropped flood plain: Hedge hydraulics. Agricultural Water Management. 1996b;**31**:91-104

[30] Truong P, Van TT, Pinners E. The Vetiver system for agriculture. The First National Indian Vetiver Workshop. The Vetiver Network International Publication; 2008, Kochi, India.

[31] Rao KPC, Cogle AL, Srivastava KL. ICPdSAT Annual Report. Andhra Pradesh, India; 1992

[32] Oshunsanya SO, Are KS, Babalola O. Soil sediment accumulation and crop yields as affected by Vetiver buffer strip spacing in Southwest Nigeria. In: Proceedings of the 34th Annual Conference of Soil Science Society of Nigeria, Institute of Agricultural Research and Training, Ibadan, Nigeria; 2010. pp. 224-231

[33] Oshunsanya SO. Spacing effects of vetiver grass (Vetiveria nigritana Stapf) hedgerows on soil accumulation and yields of maize-cassava intercropping system in Southwest Nigeria. Catena. 2013;**104**:120-126

[34] Rodriguez OD. Vetiver grass technology for soil conservation on steep agricultural land. In: Proceedings of the International Workshop on Soil Erosion Processes on Steep Lands; Merida, Venezuela; 1993

[35] Truong PN, Baker D. The role of vetiver grass in the rehabilitation of toxic and contaminated lands in Australia. In: International Vetiver Workshop, Fuzhou, China; October 1997

[36] Truong, P. Vetiver system for prevention and treatment of contaminated land and water. In: Proceeding of the Fourth International Vetiver Conference, Caracas, Venezuela, 2006; p. 16

[37] Truong PN, Baker D. Vetiver grass for the stabilization and rehabilitation of acid sulphate soils. In: Proceedings of the 2nd National Conference on Acid Sulphate Soils, Coffs Harbour, Australia; 1996. pp. 196-198

[38] Techapinyawat S, Sripen K, Komkriss T. Allelopathic effects of vetiver grass on weeds. In: Abstracts of Papers Presented at ICV-1, ORDPB, Bangkok; 1996. p. 45

[39] Truong PNV. Vetiver grass technology for mine tailings rehabilitation. In: Barker D, Watson A, Sompatpanit S, Northcut B, Maglinao A. Ground and Water Bioengineering for Erosion Control and Slope Stabilisation. NH, USA: Science Publishers Inc.; 2004

[40] Truong P, Smeal C. Research, development and implementation of Vetiver system for wastewater treatment: GELITA Australia. Technical Bulletin No. 2003/3. Pacific Rim Vetiver Network. Bangkok, Thailand: Office of the Royal Development Projects Board; 2003

[41] Truong PN, Hart B. Vetiver system for wastewater treatment. Technical Bulletin No. 2001/2. Pacific Rim Vetiver Network. Bangkok, Thailand: Office of the Royal Development Projects Board; 2001

[42] Percy I, Truong P. Landfill leachate disposal with irrigated Vetiver grass. In: Proceedings of Landfill 2005. National Conference on Landfill; Brisbane, Australia; September 2005

[43] Smeal C, Hackett M, Truong P. Vetiver system for industrial wastewater treatment in Queensland, Australia. In: Proceedings of the Third International Vetiver Conference; Guangzhou, China, October 2003

[44] Wagner S, Truong P, Vieritz A, Smeal C. Response of vetiver grass to extreme nitrogen and phosphorus supply. In: Proceedings of the Third International Vetiver Conference; Guangzhou, China; October 2003

[45] Xia HP, Ao HX, Lui SZ, He DQ. A preliminary study on vetiver's purification for garbage leachate. In: Paper Presented at the International Vetiver Workshop; Fuzhou China; October 1997

[46] Zheng CR, Tu C, Chen HM. Preliminary experiment on purification of eutrophic water with vetiver. In: Paper Presented at the International Vetiver Workshop; Fuzhou, China; October 1997

[47] Chomchalow N. Review and update of the Vetiver system R&D in Thailand. In: Proceedings of the Regional Vetiver Conference; Cantho, Vietnam; 2006

[48] Liao X, Shiming L, Yinbao W, Zhisan W. Studies on the abilities of Vetiveria zizanioides and Cyperus alternifolius for pig farm wastewater treatment. In: Proceedings of the Third International Vetiver Conference; Guangzhou, China; October 2003

[49] Le Van Du, Truong, P. Vetiver system for erosion control on severe acid sulfate soil in Southern Vietnam. In: Proceedings of the Third International Vetiver Conference; Guangzhou, China; October 2003

[50] Babalola O, Oshunsanya SO, Are K. Effects of vetiver grass (Vetiveria nigritana) strips, vetiver grass mulch and an organomineral fertilizer on soil, water and nutrient losses and maize (*Zea mays* L.) yields. Soil and Tillage Research. 2007;**96**:6-18

[51] Laing DR. Vetiver trials at the International Centre for Tropical Agriculture (CIAT), Columbia. Vetiver Newsletter. 1992;**8**:13-15